THE ENERGY ROADMAP

A Water and Wastewater Utility Guide to More Sustainable Energy Management

2013

Water Environment Federation
601 Wythe Street
Alexandria, VA 22314–1994 USA
http://www.wef.org

The Energy Roadmap:
Water and Wastewater Utility Guide to More Sustainable Energy Management

ISBN 978-1-57278-273-0

About WEF

Founded in 1928, the Water Environment Federation (WEF) is a not-for-profit technical and educational organization of 36,000 individual members and 75 affiliated Member Associations representing water quality professionals around the world. WEF members, Member Associations and staff proudly work to achieve our mission to provide bold leadership, champion innovation, connect water professionals, and leverage knowledge to support clean and safe water worldwide. To learn more, visit www.wef.org.

Contents

Acknowledgments

Special thanks are provided to the original participants in the invitation-only landmark energy summit of water and power industry leaders convened by the Water Environment Federation in Raleigh, North Carolina, in March 2012:

Mohammed M. Abu-Orf, AECOM, Philadelphia, Pennsylvania

Matt Bond, P.E., Black & Veatch, Dallas, Texas

Charles B. Bott, Ph.D., P.E., BCEE, Hampton Roads Sanitation District, Virginia Beach, Virginia

Jeanette Brown, P.E., DEE, D. WRE, University of Connecticut, Storrs-Mansfield, Connecticut

Peter Burrowes, CH2M HILL Canada Ltd., Kitchener, Ontario, Canada

Joseph C. Cantwell, P.E., Science Applications International Corporation, Brookfield, Wisconsin

Peter V. Cavagnaro, P.E., BCEE, Johnson Controls, Inc., Milwaukee, Wisconsin

Kartik Chandran, Ph.D., Columbia University, New York, New York

Lee E. Ferrell, P.E., BCEE, CEM, Schneider Electric, Anderson, South Carolina

Lauren Fillmore, Water Environment Research Foundation, Alexandria, Virginia

Anthony Fiore, New York City Environmental Protection

Rob Greenwood, Facilitator, Ross Strategic, Seattle, Washington

Jim Horne, Office of Wastewater Management, U.S. Environmental Protection Agency, Washington, D.C.

Michael P. Keleman, InSinkErator, Racine, Wisconsin

Barry L. Liner, Ph.D., P.E., Water Environment Federation, Alexandria, Virginia

Robert E. Lonergan, P.E., Tetra Tech, Denver, Colorado

Edward H. McCormick, P.E., East Bay Municipal Utility District, Oakland, California

Austin Montgomery, Carnegie Mellon Software Engineering Institute, Pittsburgh, Pennsylvania

Thomas G. Mossinger, P.E., Carollo Engineers, Inc., Walnut Creek, California

Sudhir N. Murthy, DC Water, Washington, D.C.

Kathleen M. O'Connor, P.E., New York State Energy Research and Development Authority, Albany, New York

Eileen J. O'Neill, Ph.D., Water Environment Federation, Alexandria, Virginia

Robert E. Ostapczuk, Pirnie/Arcadis, Clifton Park, New York

Coert P. Petri, MSc, Waterboard Rijn en Ijssel, Netherlands

Matthew P. Ries, P.E., Water Environment Federation, Alexandria, Virginia

Maurice Rousso, Heliopower Inc., Murrieta, California

Ralph B. "Rusty" Schroedel, Jr., P.E., BCEE, Brown and Caldwell, Milwaukee, Wisconsin

Stephen Tarallo, Black & Veatch, Gaithersburg, Maryland

Kristina Twigg, Water Environment Federation, Alexandria, Virginia

Jason Turgeon, U.S. Environmental Protection Agency, Boston, Massachusetts

Art K. Umble, Ph.D., P.E., BCEE, MWH Americas, Inc., Denver, Colorado

Scott R. Vandenburgh, CDM Smith, Seattle, Washington

John L. Willis, P.E., Brown and Caldwell, Atlanta, Georgia

Development of this document was led by Edward H. McCormick, P.E., and Alicia R. Chakrabarti, P.E., of East Bay Municipal Utility District in Oakland, California. Technical oversight was also provided by Barry L. Liner, Ph.D., P.E., of the Water Environment Federation.

Authors of the document are as follows:

Executive Summary

Alicia R. Chakrabarti, P.E., East Bay Municipal Utility District, Oakland, California

Edward H. McCormick, P.E., East Bay Municipal Utility District, Oakland, California

1 Introduction

Alicia R. Chakrabarti, P.E., East Bay Municipal Utility District, Oakland, California

Edward H. McCormick, P.E., East Bay Municipal Utility District, Oakland, California

Tania Datta, Ph.D., Tennessee Tech University, Cookeville, Tennessee

Robert H. Forbes, Jr., CH2M HILL, Inc., Charlotte, North Carolina

David L. Parry, CDM Smith, Bellevue, Washington

Jason Turgeon, U.S. Environmental Protection Agency, Boston, Massachusetts

Drury Denver Whitlock, P.E., CH2M HILL, Salt Lake City, Utah

2 Strategic Management

Chris J. Peot, DC Water, Washington, D.C.

Jim Horne, Office of Wastewater Management, U.S. Environmental Protection Agency, Washington, D.C.

Milind Wable, Ph.D., P.E., BCEE, Oceanside, California

John Wright, Raftelis Financial Consultants, Inc., Kansas City, Missouri

3   Organizational Culture

Karen L. Pallansch, Alexandria Renew Enterprises (ARenew), Alexandria, Virginia

Joseph C. Cantwell, P.E., Science Applications International Corporation, Brookfield, Wisconsin

Paul M. Kohl, P.E., Philadelphia Water Department (PWD), Philadelphia, Pennsylvania

4   Communication and Outreach

Jason Turgeon, U.S. Environmental Protection Agency, Boston, Massachusetts

Alicia R. Chakrabarti, P.E., East Bay Municipal Utility District, Oakland, California

Craig Miller, Ph.D., CDM Smith, Fortitude Valley, Queensland, Australia

Samantha Villegas, SaVi PR, LLC, South Riding, Virginia

5   Demand-Side Management

David J. Reardon, P.E., BCEE, ENV SP, HDR Engineering, Inc., Folsom, California

Tania Datta, Ph.D., Tennessee Tech University, Cookeville, Tennessee

Robert E. Lonergan, P.E., Tetra Tech, Denver, Colorado

Ralph "Rusty" B. Schroedel, Jr., P.E., BCEE, Brown and Caldwell, Milwaukee, Wisconsin

Vamsi Seeta, Atkins North America, Glendora, California

William E. Toffey, Effluential Synergies, LLC, Philadelphia, Pennsylvania

Drury Denver Whitlock, P.E., CH2M HILL, Salt Lake City, Utah

6   Energy Generation

Ralph "Rusty" B. Schroedel, Jr., P.E., BCEE, Brown and Caldwell, Milwaukee, Wisconsin

Tania Datta, Ph.D., Tennessee Tech University, Cookeville, Tennessee

Robert E. Lonergan, P.E., Tetra Tech, Denver, Colorado

David J. Reardon, P.E., BCEE, ENV SP, HDR Engineering, Inc., Folsom, California

Maurice Rousso, Heliopower Inc., Murrieta, California

Vamsi Seeta, Atkins North America, Glendora, California

Anthony Tartaglione, M.S., P.E., Treatment IV Wastewater Operator, DOE Qualified Pumping Systems Specialist, North American Board of Certified Energy Practitioners, PV Entry Level Certificate of Knowledge, Denville, New Jersey

Drury Denver Whitlock, P.E., CH2M HILL, Salt Lake City, Utah

Patrick Wootton, P. E., Nixon Energy Solutions, Lawrenceville, Georgia

7 Innovating for the Future

William E. Toffey, Effluential Synergies, LLC, Philadelphia, Pennsylvania

Flor (June) Garcia Becerra, CH2M HILL Canada Limited, Toronto, Ontario, Canada

Wendell O. Khunjar, Ph.D., Hazen and Sawyer P.C., Fairfax, Virginia

8 Conclusions

Alicia Chakrabarti, P.E., East Bay Municipal Utility District, Oakland, California

Edward H. McCormick, P.E., East Bay Municipal Utility District, Oakland, California

9 Case Studies

Alicia R. Chakrabarti, P.E., East Bay Municipal Utility District. Oakland, California

Jeremy Cramer, City of Stevens Point Wastewater Treatment Plant, Stevens Point, Wisconsin

Alan L. Grooms, P.E., Madison Metropolitan Sewerage District, Madison, Wisconsin

Angela M. Hintz, P.E., CEM, CEA, ARCADIS, Buffalo, New York (Elmira, New York)

Jim McCaughey, Narragansett Bay Commission, Providence, Rhode Island

James P. McQuarrie, P.E., Metro Wastewater Reclamation District, Denver, Colorado

James J. Newton, P.E., BCEE, Kent County Regional Wastewater Treatment Facility, Milford, Delaware

Dan Roberts, P.E., City of Palm Bay, Florida

John Hulett, Jennifer Damon, Shawnee Dunagan, Ashley Kaiser, South Truckee Meadows Water Reclamation Facility, Reno, Nevada

Barry Wenskowicz, Narragansett Bay Commission, Providence, Rhode Island

Appendix

The Energy Roadmap: A Water and Wastewater Utility Guide to More Sustainable Energy Management and ISO 50001—Energy Management

Barry L. Liner, Ph.D., P.E., Water Environment Federation, Alexandria, Virginia

Additional reviewers of the document include

Richard G. Atoulikian, PMP, P.E., MWH Americas, Inc., Strongsville, Ohio

Peter V. Cavagnaro, P.E., BCEE, Johnson Controls, Inc., Milwaukee, Wisconsin

Lauren Fillmore, Water Environment Research Foundation, Alexandria, Virginia

Richard E. Finger, Kent, Washington

Ely Greenberg, P.E., CEM, Erg Process Energy, New York, New York

Angela M. Hintz, P.E., CEM, CEA, ARCADIS, Buffalo, New York

Michael Kiparsky, Ph.D., University of California, Berkeley, California

James J. Newton, P.E., BCEE, Kent County Department of Public Works, Milford, Delaware

Amanda L. Poole, P.E., Baxter & Woodman, Inc., Chicago, Illinois

Velmurugan Subramanian, Dewberry, Fairfax, Virginia

Harry Zhang, Ph.D., P.E., CH2M HILL, Chantilly, Virginia

Executive Summary

"A renewable energy revolution is underway…and this 'sea change' is rapidly transforming the water sector landscape from treatment of water alone to the recovery of valuable resources—clean water, vital nutrients, and renewable energy. *The Energy Roadmap: A Water and Wastewater Utility Guide to More Sustainable Energy Management* will help to innovate and cost-effectively turn treatment facilities into 'green factories'".

Edward H. McCormick, Manager of Wastewater Engineering,
East Bay Municipal Utility District, Oakland, California

In early 2012, a group of water and energy industry leaders convened by the Water Environment Federation identified the need for an energy guidance document to guide water utilities of all sizes on the path to sustainable energy management, including increased renewable energy production, energy conservation, and a steady, long-term focus on energy management. While it may not be practical for all utilities to become energy-positive or neutral, all utility managers can take significant steps toward increasing energy sustainability. The resulting *The Energy Roadmap: A Water and Wastewater Utility Guide to More Sustainable Energy Management* (*The Energy Roadmap*) presents guidance for all "water" utilities (here, the term, *water,* is intended to be inclusive of drinking water and wastewater).

Organization of this Guidance Document

The Energy Roadmap presents six interrelated energy management topic areas:

- Strategic Management (Section 2),
- Organizational Culture (Section 3),
- Communication and Outreach (Section 4),
- Demand-Side Management (Section 5),
- Energy Generation (Section 6), and
- Innovating for the Future (Section 7).

1

For each topic area, there are three levels of progression (from "enable" to "integrate" to "optimize" [see Figure ES.1]). A utility may choose to focus on one or two topic areas and, over time, may eventually address three or more to further advance its energy program. Utility managers should choose which topic areas to focus on and what level of progression they wish to achieve. *The Energy Roadmap* provides practical approaches to help utility managers toward their goals, including prioritizing actions to take toward energy sustainability. For example,

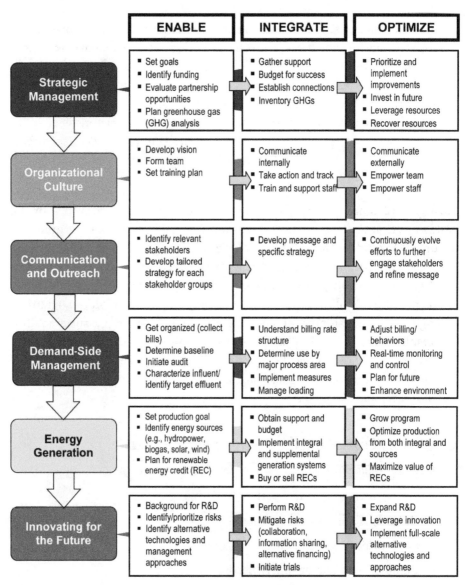

FIGURE ES.1   *The Energy Roadmap: A Water and Wastewater Utility Guide to More Sustainable Energy Management* framework (R&D = research and development).

many small utilities, particularly those without an anaerobic digestion process, may find that focusing on demand reduction (rather than energy generation) will yield a better "return on investment" of their energy management dollar. It is important to keep in mind that there is no one model for all utilities.

Topic Areas Addressed

Strategic Management

It is important to start with clear and specific energy goals and integrate them to a utility's strategic planning and policy-setting process. Goals should be used to create key performance indicators (KPIs) that can be measured and tracked. Accurate and effective branding through policy statements conveys the benefits of this focus on sustainable operations to the public and policymakers.

Energy demand reduction and renewable energy production projects support bottom-line financial sustainability by creating cost savings or revenue generation; however, some benefits of energy projects can be more clearly identified using more holistic accounting methods, such as life cycle cost analysis (LCA). Energy projects may also benefit from creative collaboration outside of the water sector. For example, a utility manager may consider working with an energy services company that can provide both technical expertise and financing assistance.

Finally, producing renewable energy and reducing energy use are both means to reducing a utility's carbon footprint. As regulation of greenhouse gases

---

Potential energy goals.

☑ 75% energy produced on site

☑ 20% energy conservation

☑ 90% energy used from renewable resources

☑ 50% of employees using public transportation

☑ 25% reduction in utility energy footprint

☑ 30% reduction in greenhouse gas emissions

Collaborate outside the sector.

During the 1990s, when Southwest Airlines wanted to improve their airplane turnaround time between flights, they did not benchmark against other airlines. Instead, they studied NASCAR pit crews and developed one of the top industry programs for efficient airplane maintenance and industry-leading on-time service to customers.

becomes more prevalent, these energy projects will become more attractive to a utility's bottom line and reputation for maintaining reasonable rates and acting as an environmental steward.

Organizational Culture

Utility managers should use a utility's energy goals and policy to define an "energy vision" that can be integrated throughout the organization. Specific, supporting, measurable actions should be included in each employee's performance plan. Staff involvement and investment are best achieved when a cross-functional "energy team" is established and an individual is designated to be an "energy champion" responsible for a variety of energy management tasks. It is recommended that the energy champion act as a single point of responsibility and accountability for an energy program. In addition to being a strong project manager, this individual must be respected throughout the organization; enthusiastic and knowledgeable about energy management and the treatment process; and explicitly granted the time, resources, and authority to lead the energy team to implement change. Additionally, the energy team should be accountable to utility leadership with regularly scheduled reporting.

Training for engineers, operators, and managers is often beneficial to effectively manage energy use and production. Finally, promoting industry involvement and rewarding energy conservation further enhances development of a culture that values energy management, conservation, and innovation. Therefore, utility managers should consider incentives such as robust performance planning linked to performance pay, team rewards, and organizational recognition.

Role of energy champion.

- ☑ Collect energy data
- ☑ Archive energy data
- ☑ Manage KPIs
- ☑ Facilitate energy team meetings
- ☑ Conduct energy measurement
- ☑ Verify energy results
- ☑ Monitor energy bills
- ☑ Capital funding prioritization
- ☑ Define energy procurement language
- ☑ Report on progress toward goals
- ☑ Continual program improvement

> **Energy educational elements.**
>
> ☑ Understanding a utility's energy bill
> ☑ On-peak and off-peak energy use
> ☑ Defining *peak demand*
> ☑ Operational effects to energy demand
> ☑ Identifying energy efficiency measures
> ☑ Effects of design flexibility on energy management
> ☑ Understanding equipment performance curves
> ☑ Understanding and application of LCA

Communication and Outreach

Successful projects often require coordination with, and support from, a variety of stakeholders. Indeed, it is important to proactively reach out to the stakeholders that are likely to be most affected by an energy program. Key stakeholder groups typically include one or more of the following external entities:

- Customers and community members,
- Regulatory and legislative staff,
- Media representatives,
- Environmental advocacy groups, and
- Water industry professionals.

Each group requires slightly different outreach approaches. For example, for customers and community members, it is important to highlight community benefits of a utility's energy projects. Depending on project needs and stakeholder expectations, outreach efforts can be as simple as posting information on a utility Web site or offering plant tours, or as complex as developing a media campaign complete with sound bites, videos, and targeted press releases. Working with the media can be the best way to get a message out to the widest audience; however, it is critical that a project is presented in the best light.

> **Energy project community benefits.**
>
> ☑ Local jobs
> ☑ Long-term rate benefit
> ☑ Availability of external funds to support energy generation and efficiency projects
> ☑ New commercial opportunities

Media communication tips.

☑ Be responsive (media tend to work on tight deadlines)

☑ Be open and transparent (it is okay to say "I do not know the answer and will get back to you")

☑ Stay on message (do not get pulled into an area that you are not prepared to discuss)

Demand-Side Management

Often, the most cost-effective efforts are simple operational changes that can significantly reduce energy use. There are several steps that can be taken to identify such measures. One practical approach is to start by analyzing the energy bill. For example, utility managers should determine whether the current electrical utility rate schedule best meets a utility's needs and analyze whether demand charges or energy use charges represent a significant component of costs. Key elements of an energy bill include the billing period and rate schedule, demand charges, energy charges and time of use, and total electricity costs.

The next step in identifying potential energy savings is to understand a utility's current energy use. Ideally, submetering data are available for each significant use or process area and can be compared to those used by other similar utilities through process benchmarking. The next important step is to perform an energy audit to identify potential energy savings measures through operational or capital improvements.

For future utilities, there are many opportunities to reduce total energy use by making strategic decisions in the planning and design processes. For example, utility managers should consider the measures listed in Table ES.1.

Finally, understanding and controlling wastewater sources allows utility managers to make appropriate plant modifications to ensure that treatment infrastructure is appropriately sized and configured to match actual demands.

Methods to reduce energy use.

☑ Shutting down

☑ Operate part time

☑ Operate with variable speed

☑ Operate at lower flows

☑ Operate at lower pressures

☑ Replace with more efficient equipment

TABLE ES.1   Opportunities for energy savings for future facilities.

Opportunity	Approach
Optimize diversion of solids from aerobic systems (decreasing energy demand) to anaerobic digesters (increasing energy production).	• Optimize primary clarifier design and hydraulic loading. • Add chemicals to enhance primary clarification • Use dissolved air flotation. • Treat raw wastewater or settled primary effluent with short contact process activated sludge to enhance clarification.
Minimize aeration system energy demand.	• Specify/install aeration systems with energy-efficient blowers. • Use deep aeration systems with fine-bubble diffusers. • Consider merging technologies such as ultra-high-efficiency strip aeration. • Install automated dissolved oxygen control via instrumentation. • Consider swing zones that can be used for either anoxic or aerobic treatment depending on loading conditions.
Use equipment that varies output.	• Use variable frequency drives for efficient treatment throughout diurnal flow and loading-rate variation.
Size equipment and processes appropriately.	• Consider shorter planning horizons and more frequent capacity expansion to ensure equipment and processes are used in an energy-efficient manner. • Evaluate existing plant flow and load conditions as compared to design conditions.
Optimize plant hydraulic profile.	• Develop a hydraulic design so that pumping is minimized.
Equalize plant flow or shave system peak to minimize flow/load variation.	• Consider collection system equalization via smart grid technology or larger systems such as deep tunnel storage to minimize peaking events that lead to oversizing of equipment/systems.
Use automation/supervisory control and data acquisition/computer technology to minimize energy use.	• Use instruments and automation appropriately to optimize system efficiency. • Consider having an energy dashboard integrated to the SCADA system to monitor and record energy metrics. • Use "real-time" measurement of demand to optimize processes and minimize energy use.
Consider auxiliary process energy use and overall environmental effect.	• Consider key processes such as odor control systems, water reuse, and tertiary treatment to meet low effluent limits and disinfection.

Energy Generation

Investing in on-site renewable energy achieves multiple goals, including increased operational reliability, reduced power costs and increased power revenue, and reduced operating cost variability linked to energy prices. It also helps to reduce a utility's carbon footprint by reducing fossil fuel consumption. Developing an energy generation strategy requires developing goals, assessing and overcoming regulatory hurdles, and identifying and prioritizing available energy resources. Energy may be derived from inherent sources within raw water, drinking water, or wastewater (e.g., hydropower, biogas, and inherent heat) or external, supplemental sources (e.g., co-digestion, solar, and wind). For the past several decades, many utilities have produced significant renewable energy from biogas. Section 6 includes a table documenting wastewater facilities in North America that are cost-effectively generating (or plan to) renewable energy using biogas. However, thousands of facilities in North America continue to flare their valuable biogas, which represents a significant opportunity. There are a number of complex considerations that enter into the decision of how best to use available biogas (see Figure ES.2). The following guidance documents and resources are also recommended:

- *Life Cycle Assessment Manager for Energy Recovery (LCAMER)* (WERF, 2012);

- *Evaluation of Combined Heat and Power Technologies for Wastewater Treatment Facilities* (Wiser et al., 2012);

- *Opportunities for Combined Heat and Power at Wastewater Treatment Facilities: Market Analysis and Lessons from the Field* (U.S. EPA, 2011); and

- U.S. Environmental Protection Agency Combined Heat and Power Web site (http://www.epa.gov/chp/).

In addition to biogas, other means to capture inherent energy in water include established technologies such as hydroelectric power and heat pumps and emerging technologies such as microbial fuel cells, biosolids gasification, and production of algae-based fuels. With advances in technology, smaller amounts of kinetic energy may be cost-effectively harvested using "in-conduit" (or "in-channel") turbines from sources such as drinking water conveyance pipelines or wastewater outfalls.

A utility can also generate energy using supplemental renewable energy sources such as co-digestion of high-strength organic wastes, solar electric, solar thermal, wind power, and other emerging technologies. As solar (photovoltaic) technologies have matured, costs are now a fraction of what they were a couple of decades ago. Utilities are often well positioned to use existing land for solar and wind projects. A key driver is the increasing prevalence and value of renewable energy certificates and renewable portfolio standards regulations.

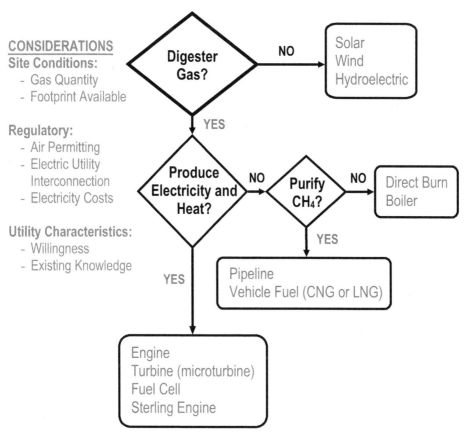

FIGURE ES-2   Energy generation decision tree (CNG = compressed natural gas and LNG = liquefied natural gas).

Innovating for the Future

It is essential that the water sector focus on the newest frontiers of water treatment and energy generation. One of the monumental challenges to the water sector is the need to create a much more energy-efficient secondary treatment "gold standard" to replace the 100-year-old activated sludge process. Several researchers are focusing their efforts on achieving full-scale anaerobic secondary treatment. Emerging technologies hold tremendous promise for further revolutionizing the water treatment industry. Significant changes require new mindsets and new practices that inherently carry risk. For smaller utilities, the first step is to understand existing and emerging technologies and the important role for public and private "water innovation hubs" and academic organizations in advancing these technologies and reducing the risk to individual utilities. Risks associated with trying and implementing new technologies include

- Technology,
- Financial,

- Regulatory, and
- Market.

Risks can be reduced through collaborative research and information sharing, innovative procurement procedures, and comprehensive sectorwide information sharing. Larger utilities not only have the ability, but also the necessity, to perform in-house research to develop and demonstrate technologies and concepts of the future. There are a number of emerging innovative technologies that are currently being developed, some of which are likely to become the standard in coming decades. Innovative, emerging technologies include low-energy treatment techniques (e.g., anaerobic secondary treatment, low-energy biological ammonia removal, and microbial fuel cells) and energy-producing treatment approaches (e.g., thermochemical conversion of biosolids to electricity or fuel; algae for biofuels; and fats, oils, and grease to biodiesel).

Case Studies

Section 9 presents two types of case studies. "Global Successes" (or retrospective) case studies provide descriptions of utilities with well-developed energy programs and how their approaches mirror the ideas and methods described in *The Energy Roadmap*. "Test Drives" (or planning) case studies provide descriptions of utilities that will be using *The Energy Roadmap* to develop and enhance their energy programs.

References

U.S. Environmental Protection Agency (2011) *Opportunities for Combined Heat and Power at Wastewater Treatment Facilities: Market Analysis and Lessons from the Field*; U.S. Environmental Protection Agency, Combined Heat and Power Partnership: Washington, D.C.

Water Environment Research Foundation (2012) *Life Cycle Assessment Manager for Energy Recovery (LCAMER)*; Water Environment Research Foundation: Alexandria, Virginia.

Wiser, J.; Schettler, J.; Willis, J. (2012) *Evaluation of Combined Heat and Power Technologies for Wastewater Treatment Facilities*; EPA-832/R-10-006; Prepared for Columbus Water Works, Columbus, Georgia; Brown and Caldwell: Atlanta, Georgia.

Section 1

Introduction

"Over the coming decades, our industry will transform itself from a waste-centered business to a resource-production business with energy at the core".

Alicia Chakrabarti, Wastewater and Energy Engineer,
East Bay Municipal Utility District, Oakland, California

In the early 20th century in North America, wastewater treatment was revolution-ized with the widespread adoption of centralized treatment. Half a century later, the U.S. Environmental Protection Agency (U.S. EPA) was formed and the Clean Water Act became law, leading to widespread implementation of secondary treat-ment of wastewater. Now, during the 21st century, the industry is on the leading edge of similar sweeping changes as utility managers say goodbye to central-ized wastewater treatment as a resource-intensive process that consumes energy, chemicals, and land to produce effluent for discharge to waterbodies. Indeed, the industry is now beginning to explore innovative approaches to cost-effectively recover and re-use resources by combining natural processes with cutting-edge technological innovations to transform materials once called "wastes" to valuable products such as clean water, vital nutrients, and renewable energy.

Some of the nation's forward-looking utilities have already begun these changes (see Section 9 for case studies) and others are prepared to follow. Industry leaders believe that considerable progress will be made toward these goals in the next 20 years, dramatically changing current standard practices that may seem archaic in retrospect. It is also believed that all utilities will significantly improve their energy management, reduce their energy use, and generate additional renewable energy. While utilities may first embark on this journey because it is fiscally prudent, they will also benefit from increased stakeholder support because it is the "right" thing to do for the environment. Furthermore, it is believed that, within 20 years, the industry will essentially eliminate flaring of biogas and reduce secondary treat-ment energy use by at least 20% while improving the quality of end products.

What's Driving Energy Sustainability?

Resource limitations, financial constraints, customer expectations, increased value of renewable energy, climate change, and surging interest in environmental sustainability are driving the transformation of treatment plants to resource recovery facilities. There are several water resource recovery facilities worldwide that have met the goal of producing more electricity than they consume and many more are nearing this milestone. To continue the industry's forward trajectory, the Water Environment Federation (WEF) and industry leaders identified the need for an energy guidance document for all water utilities (here, the term, *water,* is intended to be inclusive of both drinking water and wastewater activities) of all sizes as they progress in this direction. Hence, *The Energy Roadmap: A Water and Wastewater Utility Guide to More Sustainable Energy Management* was created. While it is not practical for all facilities to become energy-positive or neutral, all utility managers can take steps toward increasing energy sustainability to benefit their customers.

Energy Management Matrices

The core of *The Energy Roadmap* is the energy management matrices (or matrices), which consist of six tables, one for each energy management topic area. Each table contains utility characteristics organized by theme, progressing from "enable" to "integrate" to "optimize" (see Tables 1.1 through 1.6). The basis of the content of the matrices originated at an invitation-only energy summit of water and power industry leaders convened by WEF in Raleigh, North Carolina, in March 2012. The results of the energy summit discussion were organized into the framework developed in the electric power sector to move to "smart grid" technology (Software Engineering Institute, 2011). Since then, a dedicated group of industry leaders with a broad variety of expertise has refined the matrices.

Following the Guidance Document

On average, the energy content of wastewater (chemical, hydraulic, and thermal) is far greater than the energy required to treat it (Tchobanoglous et al., 2009); energy-neutral wastewater facilities are achievable, but few to date have attained this goal.

The six matrices provide the framework for *The Energy Roadmap* (see Figure ES.1); Sections 2 through 7 provide a narrative elaboration of the six matrix topic areas. Each theme (or row within the matrix) is a subsection of *The Energy Roadmap*. The utility characteristics from the matrix are presented in graphical form at the start of each subsection to show the progression from the initial planning steps ("enable") to implementation ("integrate") toward expansion and continuous improvement ("optimize").

(Text continues on page 19)

TABLE 1.1 Strategic management—utility progression characteristics.

	Enable	Integrate	Optimize

Strategic direction	SET GOALS
	• Energy goals and key performance indicators are established for both conservation (see *Demand-Side Management*) and production (see *Energy Generation*).
		GATHER SUPPORT
		• Utility incorporates energy goals and key performance indicators to strategic plan.
		• Governing board establishes energy/sustainability committee.
			PRIORITIZE AND IMPLEMENT
			• Energy management program initiatives are prioritized using tools such as
			  ○ Strategic business planning and effective utility management and
			  ○ Environmental management systems.
			• Energy generation is an integral part of a utility's suite of services.
			• Utility implements ISO 50001 Standard.
			• Utility uses triple bottom line approach for sustainability project decision making.

Financial viability	IDENTIFY FUNDING OPTIONS • Financial strategy developed to support energy audit and to fund resulting projects.	BUDGET FOR SUCCESS • Life cycle analysis used for decision-making on energy projects. • Energy use is considered on all capital project design and in operating budget decisions and standard operating practices.	INVEST IN THE FUTURE • Utility's energy initiatives generate sufficient revenue to invest in other utility priorities and reduce upward pressure on rates. • Energy arbitrage opportunities are leveraged.
Collaborative partnerships	EVALUATE OPPORTUNITIES • Opportunities for collaboration on energy projects (e.g., energy services company, joint venture, public–public/private partnership) are analyzed. • Diverse markets for energy products are identified.	ESTABLISH CONNECTIONS • Contracts with partners are in place and implemented to facilitate data exchange and planning with water, energy, and gas utilities. • Utility planning efforts are integrated with other agencies regarding multiple resources (e.g., water, stormwater, etc.).	LEVERAGE RESOURCES • Utility uses partnerships to maximize energy sales revenues and/or reduce demand (e.g., selling power or biogas to adjacent facility, working with a feedstock provider for co-digestion).
Toward carbon neutrality	PLAN CARBON FOOTPRINT ANALYSIS • Approach to carbon footprint analysis/GHG inventory is established.	INVENTORY GHG* EMISSIONS • Carbon footprint/GHG inventory is developed.	RECOVER RESOURCES • Additional resources are recovered or realized (e.g., carbon credits) as utility moves toward carbon neutrality. • Comprehensive carbon footprint/GHG inventory is maintained, including fugitive emissions and embodied energy of significant inputs (e.g., chemicals).

*GHG = greenhouse gas.

13

TABLE 1.2 Organizational culture—utility progression characteristics.

	Enable	Integrate	Optimize
Energy vision	**DEVELOP VISION** • Leadership group develops energy vision. • Governing body adopts energy vision as policy. • Leadership group communicates energy vision to workforce.	**COMMUNICATE INTERNALLY** • Leadership group links energy vision to staff performance plans. • Leadership group incorporates energy goals/key performance indicators to strategic plan.	**COMMUNICATE EXTERNALLY** • Utility shares energy vision with external stakeholders and the industry. • Plans are in place to embrace external market changes.
Energy team	**FORM TEAM** • Utility establishes cross-functional energy team. • Leadership group establishes clear charge and authority for energy team with defined roles for members.	**TAKE ACTION AND TRACK** • Energy team drives implementation of recommendations. • Energy team systematically reports on progress and future actions.	**EMPOWER TEAM** • Energy team provided significant budget authority to implement improvements. • Energy team interfaces directly with governing body to get direction from and report on energy program status.
Staff development and alignment	**SET TRAINING PLAN** • Employee performance plans include energy program-related activities to support energy vision. • Training needs for utility leadership and staff are identified.	**TRAIN AND SUPPORT STAFF** • Staff are trained in demand-side management and energy generation. • Staff maintains knowledge of emerging technologies through information-sharing events.	**EMPOWER STAFF** • Leadership group establishes incentives for energy-conservation results. • Leadership group empowers staff to make changes for energy savings.

TABLE 1.3 Communication and outreach—utility progression characteristics.

	Enable	Integrate	Optimize
Customers and community	• Customer outreach and education strategy is tailored to project needs and customer expectations. • Community groups are identified for outreach to gain program support.	• Proactive customer outreach program (e.g., bill inserts, tours, fact sheets, Web site) that focuses on environmental benefits and cost-effectiveness is established.	• Utility engages customers in helping to achieve energy program goals (e.g., local grease collection).
Regulatory and legislative	• Key regulators are identified and effective working relationships are established (e.g., regulations pertaining to air and solids). • Legislative strategy is developed to enhance opportunities and minimize hurdles for energy program.	• Key regulators are educated on holistic energy/water relationship. • Utility advocates for unified regulations that address cross-media issues. • Regional collaboration with other agencies occurs (e.g., for funding or policy changes).	• Utility works with industry associations to influence regulators/legislature to create incentives to encourage efficient energy use and increase renewable energy production. • Utility influences funding agencies to prioritize energy projects in the water sector. • Regulators and utility work together to resolve cross-media issues.
Media outreach	• Media outlets are identified and strategies are developed.	• Media kit is developed (e.g., video, sound-bites, pictures, and press releases).	• Dedicated utility staff work on messaging with media.
Environmental advocacy groups	• Outreach strategy is developed to support energy projects. • Appropriate partnerships are identified.	• Utility shares energy program activities (e.g., tours, fact sheets, etc.).	• Joint programs and outreach that support the goals of both organizations are implemented.
Water sector	• Key energy staff network at local/regional industry events and information-sharing groups.	• Successes, failures, and lessons learned are shared at industry events.	• Energy staff lead industry initiatives to support sector advancements in sustainability.

15

TABLE 1.4 Demand-side management—utility progression characteristics.

	Enable	Integrate	Optimize

Electricity costs and billing	GET ORGANIZED
	• Historical electric bills are analyzed (2-plus years of data are preferred).
		UNDERSTAND THE DETAILS
		• Rate structure and billing details are understood
		  ○ Demand charges
		  ○ Energy charges, unit costs, and time of use
		  ○ Billing period
			IMPLEMENT CHANGES
			• Modifications are made to billing and/or operations to reduce costs
			  ○ New rate structure is selected and
			  ○ Loads are shifted to reduce on-peak demand charges or unit costs.

Power measurement and control	GET THE BIG PICTURE • Baseline energy use and benchmarks are determined. • Energy submetering needs are identified. • SCADA[a] systems and power monitoring capabilities are identified.	DETERMINE USE BY KEY PROCESS • Energy use by each significant unit process area is determined. • Energy use is benchmarked against similar size/type plants to identify target areas for energy reductions. • Electricity use and process data are analyzed together. • Load management (shedding/switching) is in place.	MONITOR FOR REAL-TIME CONTROL • Electricity use by significant load center is monitored in real time. • Real-time control is in place (e.g., SCADA) to measure equipment energy use and efficiency with a user-friendly display (i.e., "energy dashboard"). • Excess power generation is wheeled to other assets or entity.
Energy management	INITIATE AUDIT • Energy team performs energy audit. • Goals are set for reducing energy use and costs.	IMPLEMENT RECOMMENDATIONS • Cost-effective recommendations from audit are implemented. • Energy team tracks actual vs planned results.	PLAN FOR THE FUTURE • Energy savings are incorporated to the design of all future capital projects and new operating strategies.
Source control	UNDERSTAND INFLUENT • Loads (industrial, water use, I&I[b]) are understood and evaluated for energy treatment requirements and energy production potential.	MANAGE LOADING • Methods are in place to manage influent loading to reduce energy use (e.g., industrial surcharge optimization, I&I reduction program, etc.). • Methods to reduce flows are investigated.	ENHANCE ENVIRONMENT • Sources are managed to reduce energy use and maximize energy production potential (e.g., appropriate incentives for trucking high-strength waste).

[a]SCADA = supervisory control and data acquisition and

[b]I&I = infiltration and inflow.

TABLE 1.5   Energy generation—utility progression characteristics.

	Enable	Integrate	Optimize
Strategy	SET PRODUCTION GOAL • Measurable energy generation goal is established. • Energy generation plan is coordinated with utility strategic plan. • Energy team understands regulatory and permit limitations (e.g., air emissions) with regard to generation.	OBTAIN SUPPORT • Governing body approves capital budget for energy generation projects. • Regulatory issues have been addressed and satisfactorily resolved.	GROW PROGRAM • Infrastructure for energy generation is proactively maintained, renewed, and upgraded. • Holistic evaluation methodologies (e.g., triple bottom line) are used to evaluate energy generation opportunities.

Energy from water	EVALUATE INTEGRAL ENERGY SOURCES
	• Available energy resources are quantified, such as
	  ○ Biogas,
	  ○ Hydropower, and
	  ○ Heat.
		IMPLEMENT GENERATION SYSTEMS
		• Energy generation facilities are operating and producing power/heat for utility use
		  ○ Electricity/heat and
		  ○ Fuel (natural gas, pellets, etc.).
			OPTIMIZE PRODUCTION
			• Energy production is optimized to maximize the value of generation (e.g., biogas storage to offset power purchases during on-peak hours).

Supplemental energy sources	IDENTIFY SUPPLEMENTAL ENERGY SOURCES
	• Available non-water-derived energy sources are quantified, including
	  ○ Co-digestion,
	  ○ Solar, and
	  ○ Wind.
	• Feedstock market evaluation is performed.
		IMPLEMENT GENERATION SYSTEMS
		• Energy generation facilities are operating and producing power/heat or fuel.
		• Quantity and quality of feedstock meets capacity.
			MAXIMIZE PRODUCTION
			• On-site electricity generation from all sources approaches or exceeds on-site electricity demand.
			• High-strength organic waste (e.g., food; fats, oils, and grease; etc.) is integrated into feedstock supply to increase generation potential.

Renewable energy certificates (RECs)	PLAN FOR RECs • Staff gain understanding of state regulations for renewable portfolio standard and production and sales of RECs.	USE RECs • Utility produces, sells, and/or purchases RECs, as appropriate.	MAXIMIZE VALUE OF RECs • Sales and purchases of RECs are optimized to maximize value of resources, potentially using automation.

TABLE 1.6 Innovating for the future—utility progression characteristics.

	Enable	Integrate	Optimize
Research and development (R&D)	PREPARE FOR R&D • Staff are well versed in existing technologies. • Opportunities are identified by survey of emerging technologies.	PERFORM R&D • Utility budget includes R&D funding. • Utility actively participates in water innovation partnerships (e.g., water innovation centers, research foundations, university partnerships, etc.).	EXPAND R&D • Utility culture is open to new technologies. • Site visits to facilities using innovative technologies occur regularly. • Completed trials and research projects provide the foundation for further advancement within the industry.
Risk management	IDENTIFY AND PRIORITIZE RISKS • Risk of innovation is identified. • Strategy for risk mitigation is developed. • Planning includes measures for climate change adaptation (e.g., extreme events).	MITIGATE RISKS • Risk is reduced through collaborative research and information sharing. • Leadership group recognizes and rewards innovative approaches.	LEVERAGE INNOVATION • Organization can successfully trial and implement innovative projects and is adaptable to emerging opportunities. • Patents are obtained to protect utility and water sector.
Alternative treatment technologies	EVALUATE TECHNOLOGIES • Technologies that reduce energy use or increase generation are identified.	INITIATE TRIALS • Advanced low-energy treatment technologies and energy production technologies are demonstrated.	IMPLEMENT FULL-SCALE SOLUTION • Lower energy-consuming processes replace energy-intensive secondary treatment.
Alternative management approaches	IDENTIFY ALTERNATIVES • Decentralized treatment options are considered. • Planning is performed on a watershed basis.	IMPLEMENT ALTERNATIVES • Green infrastructure projects are implemented where appropriate. • Enhanced regionalization (e.g., biosolids processing) has been considered and implemented where appropriate.	EXPAND INTEGRATION • Alternative management approaches (e.g., decentralization, regionalization, etc.) are used, where appropriate, to maximize overall, regionwide benefit.

The Energy Roadmap is organized by topic areas (shown on the left in Figure ES.1) and levels of progression within each topic area (shown across the top in Figure ES.1). The vertical arrows between the topic areas represent how the document is organized, but are not meant to imply a linear stepwise progression from one topic to the next.

Topic Areas

The following topic areas include both organizational and technical aspects of energy management:

- Strategic management—high-level management policies, practices, and measurable goals that lay the foundation for sustainable energy management;

- Organizational culture—implementation of an energy vision to create an organizational culture that values efficient energy use at all levels and supports an energy champion and cross-functional energy team;

- Communication and outreach—tools for effective two-way communication with key stakeholders on energy management;

- Demand-side management—methods to assess and reduce energy use and energy costs, including energy bill analysis, energy audit, and real-time energy monitoring and control;

- Energy generation—tools for a utility to evaluate whether and how to increase on-site renewable energy production and/or investments, including sources inherent to water and other supplemental renewable energy sources; and

- Innovating for the future—guidance for utilities of all sizes to leverage existing research on emerging, innovative treatment technologies and approaches, further in-house innovation, and manage risk associated with these ventures.

Levels of Progression

The three levels of progression within each topic area are defined as follows:

- Level 1, Enable—describes the characteristics of the planning process, including initiating first steps and launching energy program components;

- Level 2, Integrate—describes the characteristics of the implementation process, including establishing and using an energy program framework to make widespread adoption within the utility successful; and

- Level 3, Optimize—describes the characteristics associated with further enhancing and fine-tuning energy program improvements and extending them outside of the utility.

How to Use the Guidance Document

Progression toward the utility of the future is based on a process of continuous improvement. First, utility managers should select a goal that can realistically be achieved within a 3- to 10-year timeframe. Utility managers should choose a goal that is measurable and then regularly track progress toward meeting it. A utility that is new to the field of energy management might select an initial goal of increasing energy efficiency and conservation (e.g., reduce energy use by 10% within 3 years). Conversely, a utility that has a long history of energy efficiency and that also has begun to generate its own power through anaerobic digestion might, perhaps, select a goal of producing more power than it consumes. Not all facilities can become "power positive", nor should they expect to do so.

A simple process to identify and prioritize which topic areas to focus on first is recommended. The approach uses the effective utility management prioritization process, which was developed by a collaborative team that included U.S. EPA and six national water associations (American Public Works Association, the American Water Works Association, the Association of Metropolitan Water Agencies, the National Association of Clean Water Agencies, the National Association of Water Companies, and WEF) (http://www.watereum.org). Utility managers should convene their utility leadership groups (potentially supplemented by elected officials, customers, etc.) and follow these three steps:

1. Use the matrix to determine your current level of achievement/performance (enable to optimize) in each of the six topic areas,

2. Rate the level of importance of each topic area to your utility, and

3. Prioritize topic areas by selecting and starting with the one or two that you collectively gave a low rating for current level of achievement and a high rating for importance.

Plotting the results of ratings can aid in the selection of priority areas, as shown in Figure 1.1. The topic areas falling in the shaded region are the ones to focus on first (e.g., Demand-Side Management [DSM] and Organizational Culture [OC]). Once progress has been made in these areas, the utility manager should consider working on topic areas that demonstrated the next highest priority for his or her utility (those closest to the bottom right quadrant).

It is important to set goals for future progression levels and use the narrative, case studies, and references to plan activities to advance achievement in these topic areas. Finally, following implementation, the matrices should be used to determine new progression levels and these levels should be compared to the original goals.

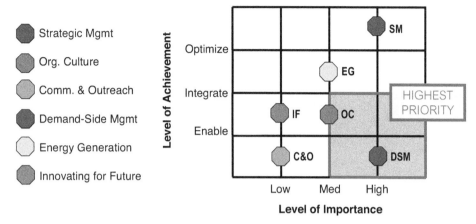

FIGURE 1.1   Importance and progression ratings to prioritize focus areas.

Scope of Coverage of the Guidance Document

Energy sustainability is one part of the overall sustainability of wastewater utilities. In the future, WEF plans to provide similar guidelines for nutrient/metal recovery and water recycling, both of which comprise important components of the industry's utility of the future. Ensuring high-quality water treatment to protect public health and the environment will continue to be the core responsibility of water resource recovery facility operators. However, protecting the environment for future generations (sustainability) requires the water sector to do even more. The water sector is in a unique position of being able to produce more renewable energy than is needed to run its facilities. Increasing energy sustainability, nutrient/metal recovery, and water recycling are quickly evolving into additional essential core practices expected of utilities.

The chief U.S. regulator of drinking water and wastewater quality, U.S. EPA, recently published *Planning for Sustainability: A Handbook for Water and Wastewater Utilities* (U.S. EPA, 2012). Regulation of carbon dioxide and other greenhouse gases (GHGs) in the United States may not be far behind. In many European countries, active GHG markets already exist and, in several U.S. states and Canadian provinces, markets are currently being developed.

Defining Energy Program Sustainability Goals

Reducing energy use and increasing energy production are not the only goals. The balance between energy efficiency and resource recovery involves trade-offs (see Figure 1.2). For example, treating wastewater to higher standards for reuse can require more energy than the potable water it displaces. Similarly, increased energy use is often required to further process biosolids to maximize reuse potential and recover nutrients and minerals (e.g., nitrogen, phosphorus,

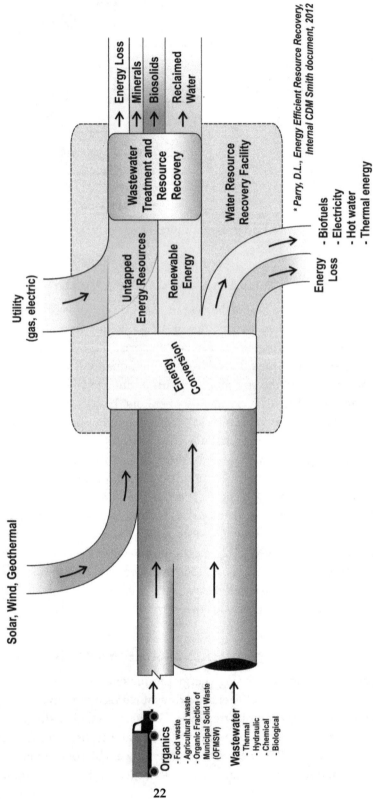

FIGURE 1.2 Schematic energy flow at a water resource recovery facility.

and magnesium). These tradeoffs must be understood and managed to achieve a utility's particular sustainability goals. In short, there is no one model that works for all utilities.

References

Software Engineering Institute (2011) Smart Grid Maturity Model: SGMM Model Definition, Version 1.2 September 2011;Technical Report CMU/SEI-2011-TR-025; Software Engineering Institute: Pittsburgh, Pennsylvania.

Tchobanoglous, G.; Leverenz, H.; Gikas, P. (2009) Impacts of New Concepts and Technology on the Energy Sustainability of Wastewater Management, Climate Change, Sustainable Development, and Renewable Energy Sources; Paper presented at Climate Change, Sustainable Development, and Renewable Energy Sources; Thessaloniki, Greece; Oct 15–17.

U.S. Environmental Protection Agency (2012) *Planning for Sustainability: A Handbook for Water and Wastewater Utilities*; EPA-832-R-12-001; U.S. Environmental Protection Agency: Washington, D.C.

Section 2

Strategic Management

"Clear strategic direction changes the tone from 'waste' to 'resource' and provides the foundation for advancing energy sustainability within the utility".

Chris Peot, Biosolids Manager, DC Water, Washington, D.C.

The strategic management topic area comprises the following four themes:

- Strategic direction,
- Financial viability,
- Collaborative partnerships, and
- Toward carbon neutrality.

Strategic Direction

ENABLE	INTEGRATE	OPTIMIZE
SET GOALS	**GATHER**	**PRIORITIZE AND IMPLEMENT**
• Energy goals and key performance indicators are established for both conservation (see Demand-Side Management) and production (see Energy Generation)	• Utility incorporates energy goals and key performance indicators to strategic plan • Governing board establishes energy/sustainability committee	• Energy management program initiatives are prioritized using tools such as ○ Strategic Business Planning ○ Effective Utility Management (EUM) ○ Environmental Management Systems (EMS) • Energy generation is an integral part of utility's suite of services • Utility utilizes triple bottom line approach for sustainability project decision making

Developing Policies

Effective strategic direction for sustainable energy management requires creation of specific energy goals. Goals should follow the "SMART" principles (i.e., specific, measurable, achievable, relevant, and time-bound). These goals can take a variety of forms and should reflect a utility's priorities and circumstances, as shown above. Policies and key performance indicators should be developed and documented in a utility's strategic plan. Another way to increase accountability and ensure that energy management is valued within a utility is for the governing body to establish an energy committee (or energy/sustainability committee) to monitor progress toward attaining goals (e.g., see the specific policy statement and goals developed by St. Peters, Missouri, below). As these goals are developed, utility managers should consider engagement of key stakeholders as discussed in Section 4.

Utility managers can use specific management tools or standards, such as ISO 50001—Energy Management (ISO, 2011), to provide structure to a utility's energy management program (refer to the Appendix for more information regarding using this standard with *The Energy Roadmap*).

Potential energy goals.

- 75% energy produced on site
- 20% energy conservation
- 90% energy used from renewable resources
- 50% of employees using public transportation
- 25% reduction in utility's energy footprint

Energy policy statement and policy goals (St. Peters, Missouri).

Policy statement: The City of St. Peters Utility Department will promote the efficient use of energy and conservation of natural resources to serve the water and wastewater treatment needs of our community in a sustainable way.

Policy goals: Improve energy efficiency continuously by establishing and implementing effective energy management programs; emphasize energy efficiency in program and facility design; emphasize energy efficiency in the procurement of equipment; encourage continuous energy conservation by employees in their daily work; and support beneficial reuse of organic and chemical byproducts of treatment operations to improve the environment.

Effective Branding

Accurate and effective branding through policy statements conveys the benefits of this new focus on sustainable operations to the public and policymakers. Water resource recovery facilities across the nation are poised to truly become "green factories" that process so-called "wastewater" into soils, fertilizers, renewable energy, and recycled water while protecting public health and re-using or returning clean water to the environment.

Financial Viability

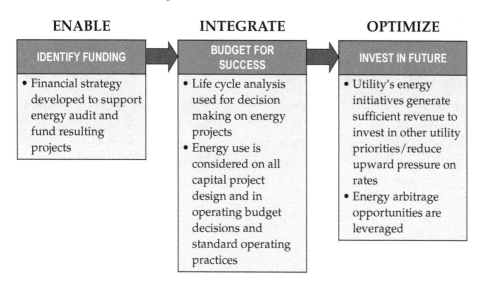

ENABLE	INTEGRATE	OPTIMIZE
IDENTIFY FUNDING	**BUDGET FOR SUCCESS**	**INVEST IN FUTURE**
• Financial strategy developed to support energy audit and fund resulting projects	• Life cycle analysis used for decision making on energy projects • Energy use is considered on all capital project design and in operating budget decisions and standard operating practices	• Utility's energy initiatives generate sufficient revenue to invest in other utility priorities/reduce upward pressure on rates • Energy arbitrage opportunities are leveraged

Rethinking Financial Metrics

Utility employees, executives, and governing bodies have a fiduciary responsibility to ratepayers and the general public when considering all investments, including technologies designed to enhance energy conservation, increase energy production, or produce products for reuse. Taking into account life cycle cost savings and triple bottom line considerations (i.e., economic, environmental, and social effects) of a new technology is important to allow for a more holistic understanding of the effects and long-term implications rather than simply considering the capital cost and simple payback. Using these economic analysis techniques can reframe the discussion of sustainable energy investments to successfully emphasize their long-term cost-effectiveness in a manner not possible using traditional metrics such as simple payback or short-term rate effects.

> Example of energy savings.
>
> A targeted energy management program in Pittsfield, Massachusetts, based on the elements in *Ensuring a Sustainable Future: An Energy Management Guidebook for Wastewater and Water Utilities* (U.S. EPA, 2008) led to energy efficiency upgrades, installation of a biomass co-generation system for on-site electric power generation, and installation of a 1575-kW solar photovoltaic system. Taken together, these projects are anticipated to result in approximately 84% savings in annual energy costs (more than $600,000) and enable the utility to generate approximately 69% of its power from renewable energy sources.

Saving Energy and Funds

In the long term, reducing energy imports (through reducing demands and/or generating on-site heat and power) will protect a utility from energy price fluctuations and continue to provide savings for many decades. Energy savings or revenues from the sale of excess green energy can then be used to reduce upward pressure on rates or reinvested in other utility priorities. In addition, on-site power production provides an additional level of resiliency from power outages, the benefits of which are significant, even if the value is not easily monetized.

Collaborative Partnerships

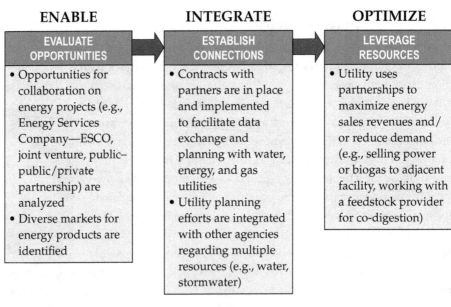

ENABLE	INTEGRATE	OPTIMIZE
EVALUATE OPPORTUNITIES	**ESTABLISH CONNECTIONS**	**LEVERAGE RESOURCES**
• Opportunities for collaboration on energy projects (e.g., Energy Services Company—ESCO, joint venture, public–public/private partnership) are analyzed • Diverse markets for energy products are identified	• Contracts with partners are in place and implemented to facilitate data exchange and planning with water, energy, and gas utilities • Utility planning efforts are integrated with other agencies regarding multiple resources (e.g., water, stormwater)	• Utility uses partnerships to maximize energy sales revenues and/or reduce demand (e.g., selling power or biogas to adjacent facility, working with a feedstock provider for co-digestion)

Partnerships for Alternative Project Funding

Utility managers can develop energy projects under traditional design–bid–build or design–build contracts, in which the utility owns and operates the asset, or they can partner with other public agencies or private companies depending on their needs and circumstances. Owning and operating a facility may be the lowest cost option in the long run, but it typically requires a significant capital outlay and carries more risk. Methods that reduce risk and capital cost include setting up a power purchase agreement or other lease-type arrangements with a developer or working with an energy services company that will fund construction or retrofit costs and be repaid over time through savings or revenues. Energy services companies offer technical expertise in energy efficiency and power generation management and capital for the investment (http://www.naesco.org).

Grant Opportunities

There are numerous grants, low-interest loans, rebates, and incentives available to provide funding assistance for renewable energy projects. For more information, the reader is referred to the Database of State and Utility Incentives for Renewables and Efficiency (http://www.dsireusa.org).

Toward Carbon Neutrality

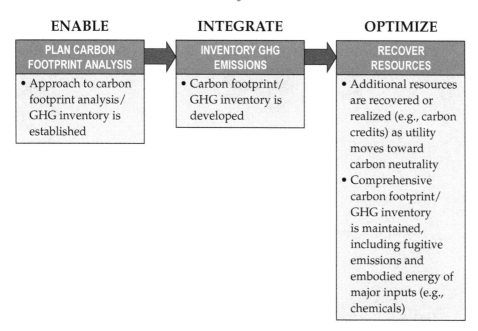

ENABLE	INTEGRATE	OPTIMIZE
PLAN CARBON FOOTPRINT ANALYSIS	**INVENTORY GHG EMISSIONS**	**RECOVER RESOURCES**
• Approach to carbon footprint analysis/ GHG inventory is established	• Carbon footprint/ GHG inventory is developed	• Additional resources are recovered or realized (e.g., carbon credits) as utility moves toward carbon neutrality • Comprehensive carbon footprint/ GHG inventory is maintained, including fugitive emissions and embodied energy of major inputs (e.g., chemicals)

As utilities look to reduce carbon footprints, consideration of all of the energy that goes into the materials being used should be evaluated. A holistic accounting of all of this energy, or the "embodied energy", includes energy associated with mining, producing, and transporting all of the resources in the product. For example, the embodied energy in chemicals used for treatment operations can be considered when evaluating plant energy performance. In addition, the utility should consider the avoided embodied energy, such as that from reduced phosphate mining when land application (or direct nutrient recovery) is used in place of chemical fertilizers.

Producing renewable energy and reducing energy use are both means to reducing a utility's carbon footprint. Many U.S. states, Canadian provinces, and countries around the world have enacted regulations that require electric providers to include a certain percentage of renewable energy within their portfolio. In the United States and Canada, this requirement is called the *renewable portfolio standards*. Some regulations also require reporting GHG emissions. In the future, states, provinces, regions, and federal governments are likely to regulate GHG emissions, which will increase the cost of energy imports and the value of renewable energy produced. These changes make the types of energy conservation and renewable energy generation projects discussed in this document all the more attractive.

References

International Organization for Standardization (2011) ISO 50001—Energy Management; International Organization for Standardization: Geneva, Switzerland.

U.S. Environmental Protection Agency (2008) *Ensuring a Sustainable Future: An Energy Management Guidebook for Wastewater and Water Utilities;* Office of Wastewater Management; U.S. Environmental Protection Agency: Washington, D.C.

Section 3

Organizational Culture

"The culture of the utility enables and drives advancements in sustainable energy management".

Karen Pallansch, General Manager, Alexandria Renew Enterprises,
Alexandria, Virginia

The organizational culture topic area comprises the following themes:

- Energy vision,
- Energy team, and
- Staff development and alignment.

Energy Vision

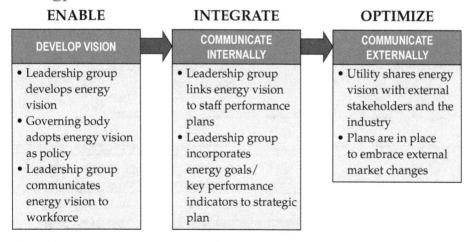

ENABLE	INTEGRATE	OPTIMIZE
DEVELOP VISION	**COMMUNICATE INTERNALLY**	**COMMUNICATE EXTERNALLY**
• Leadership group develops energy vision • Governing body adopts energy vision as policy • Leadership group communicates energy vision to workforce	• Leadership group links energy vision to staff performance plans • Leadership group incorporates energy goals/ key performance indicators to strategic plan	• Utility shares energy vision with external stakeholders and the industry • Plans are in place to embrace external market changes

Incorporating Energy Vision to Strategic Planning

Energy management becomes tangible when a utility formalizes and prioritizes an "energy vision" that is incorporated to the strategic planning process.

To make it real, use your:

☑ Strategic Plan

☑ Annual Business Plan

☑ Key Performance Indicators

☑ Employee Performance Plan

As discussed in Section 2, "Strategic Management", specific metrics need to be translated to key performance indicators (KPIs) and incorporated to a utility's annual business planning documents; similarly, they need to be included in staff performance plans (see Figure 3.1). Incorporating these measures to each team and employee performance management system and individual development plan holds every individual accountable for the overall success of the program and provides leadership with a tool to monitor progress. Vision can and should come from and be fostered at all levels within an organization. It is critical that all staff are invested in the process so that they embrace, enhance, and implement the vision.

Incorporating Energy Vision to Project Planning

Including an energy focus in individual project planning is another crucial tactic for success. Utility managers should incorporate energy-related issues as core values and/or management criteria used to make project decisions and set program priorities. Including management criteria ensures that each project or program considers energy during planning and implementation.

Using an Energy Vision to Drive Energy Goals

Presenting energy goals visually can also help make them more tangible. Figure 3.2 shows how historical data can be plotted with projected data based on established energy goals. This graph shows how a utility plans on reducing its overall energy consumption (which may include electricity, natural gas, heating oil, transportation fuel, etc.) and costs through a combination of energy efficiency and energy generation.

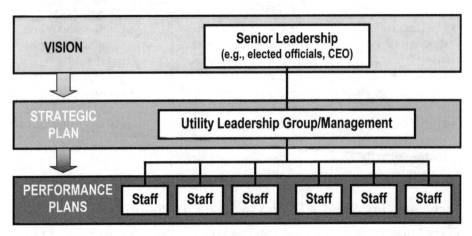

FIGURE 3.1   From vision to strategic planning and performance planning.

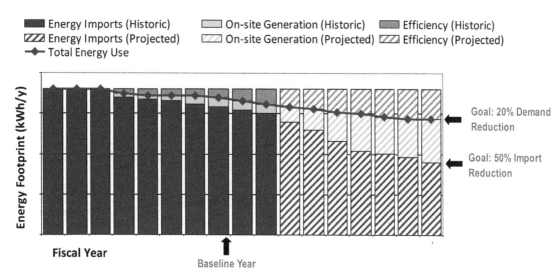

FIGURE 3.2   Development and visual presentation of energy vision and goals.

Energy Team

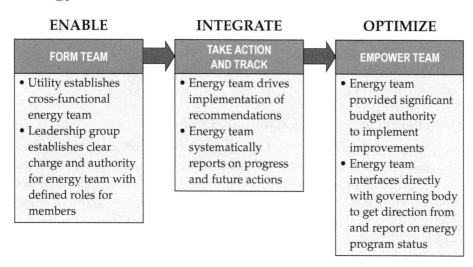

Creating an Effective Energy Team

Utility managers should establish a cross-functional energy team that engages in project planning on all aspects of projects (i.e., capital investments, operational changes, benefit allocation, and marketing). A typical energy team would include staff from engineering (civil, electrical, and control systems); operations (including process control) and maintenance (mechanical, electrical, and instrumentation); regulatory compliance; and planning (see Figure 3.3). The energy

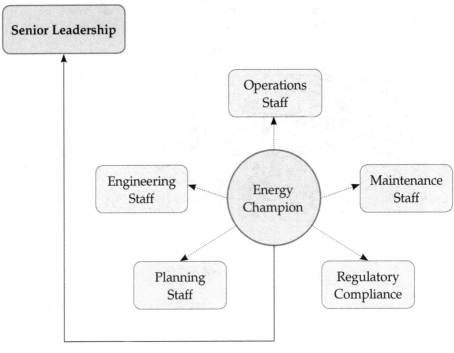

FIGURE 3.3   Possible energy team structure.

team should be accountable to and receive clear authority from the overall leadership group (i.e., the governing body or utility executives).

Designating an Energy Champion

An effective energy team is best led by a single individual or "energy champion" who can unite the various leaders, all of whom have their own primary responsibilities. This "program manager" approach ensures there is a single point of responsibility and accountability. This individual, in addition to being a strong project manager, must be respected throughout the organization, enthusiastic and knowledgeable about energy management, and must be explicitly granted the time, resources, and authority to lead the energy team and implement change. Typical tasks may include collection and archiving of energy data (and KPIs), facilitation of energy team meetings, conducting feasibility studies, and monitoring energy bills. The energy champion should also have a role in capital budgeting and project prioritization. The entire energy team should be accountable to utility leadership, reporting, on a monthly or quarterly basis, results of recent activities or studies and progress toward meeting goals.

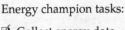

Energy champion tasks:

☑ Collect energy data

☑ Archive energy data

☑ Report and track goals

☑ Facilitate energy team meetings

☑ Conduct energy measurement

☑ Verify energy results

☑ Monitor energy bills

☑ Capital funding prioritization

☑ Define energy procurement language

Staff Development and Alignment

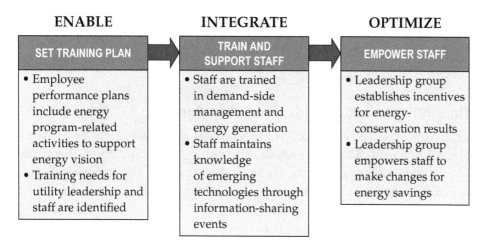

ENABLE	INTEGRATE	OPTIMIZE
SET TRAINING PLAN	**TRAIN AND SUPPORT STAFF**	**EMPOWER STAFF**
• Employee performance plans include energy program-related activities to support energy vision • Training needs for utility leadership and staff are identified	• Staff are trained in demand-side management and energy generation • Staff maintains knowledge of emerging technologies through information-sharing events	• Leadership group establishes incentives for energy-conservation results • Leadership group empowers staff to make changes for energy savings

Establishing a Training Program

Knowledge of electricity is critical to operators and maintenance staff from a safety perspective, yet energy efficiency and management are not. Although the two concepts are not the same, it is commonly believed that if you know safe electrical practices then you understand energy management. Communication of energy consumption does not necessarily flow through an organization; those responsible for paying bills and setting an energy budget may not necessarily understand how energy is used and where savings can be found. Similarly, engineers and operators who make daily choices on how energy is used often do not have the information required to understand the cost implications of their decisions.

It is recommended that a separate and succinct energy training program be developed and presented to staff associated with the operation of all aspects of

---

Energy educational elements.

☑ Understanding your energy bill

☑ On-peak and off-peak energy use

☑ Defining *peak demand*

☑ Operational effects on energy demand

☑ Identifying how to become energy efficient

☑ Effects of design flexibility on energy management

☑ Understanding equipment performance curves

☑ Understanding and use of life cycle cost analyses

a utility. Educational materials need to incorporate an explanation of the information presented on the bill, which includes a description of energy consumption and on- and off-peak demands (see Section 5 for a detailed discussion). It should also include a review of standard operating procedures to ensure that daily tasks address energy use. Procurement standards for energy efficiency should also be reviewed.

Promoting Industry Information Sharing

Commitment to ongoing success and continuous improvement depends on staying informed of the latest advances, technologies, and practices. Part of the energy team's role is to perform this scan, share insights, and recommend ideas for further investigation. Identifying key information may be accomplished through technology (e.g., electronic newsletters, monitoring Web sites, and social media), networking, and belonging to and attending water association functions.

Reinforcing and Rewarding Energy Conservation

Rewarding energy-conserving behaviors helps to make those new behaviors habits and embed them within an organization. To do this in a meaningful way, an organization must have a robust performance planning process and KPIs directly attributable to energy management outcomes. These must be monitored and reported to all employees at least quarterly. Successful, measured outcomes then support meaningful rewards, which can include anything from performance-related pay increases and spot rewards to team lunches and organizational recognition. Ingenuity in recognizing successful energy management within the various financial and ethical considerations within the industry is just as important as ingenuity in implementing research and technology for energy reductions.

Section 4

Communication and Outreach

"Effectively communicating the sustainable energy management program in a timely manner will help achieve support from stakeholders. Moreover, it's the best way to prepare them for changes, innovations, or rate increases. A solid effort will strengthen relationships with them and build or reinforce the utility's reputation as caring and responsible".

Samantha Villegas, APR, President, SaVi PR, LLC, South Riding, Virginia

The communications and outreach topic area comprises the following themes:

- Customers and community members,
- Regulatory and legislative,
- Media outreach,
- Environmental advocacy groups, and
- Water industry.

Consultation and outreach with key stakeholders, including ratepayers, regulators, legislators, environmental advocacy groups, and the media should be an active process integrated with overall strategic planning. The appropriate level of engagement will depend on the specifics of a utility's energy program and the expectations and needs of stakeholders. Therefore, utility managers should tailor their approach for each group and adjust as needed. The reader is referred to *Planning for Sustainability: A Handbook for Water and Wastewater Utilities* (U.S. EPA, 2012) for additional guidance.

Customers and Community Members

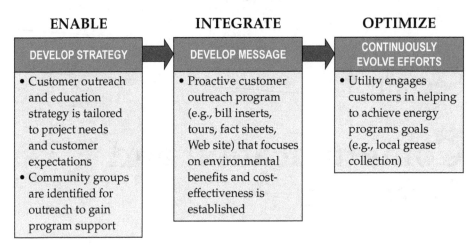

ENABLE	INTEGRATE	OPTIMIZE
DEVELOP STRATEGY	**DEVELOP MESSAGE**	**CONTINUOUSLY EVOLVE EFFORTS**
• Customer outreach and education strategy is tailored to project needs and customer expectations • Community groups are identified for outreach to gain program support	• Proactive customer outreach program (e.g., bill inserts, tours, fact sheets, Web site) that focuses on environmental benefits and cost-effectiveness is established	• Utility engages customers in helping to achieve energy programs goals (e.g., local grease collection)

Effective communication with customers and community members is critical as they can represent a utility's strongest advocate or detractor; indeed, they exert influence over elected officials that set policy for a utility and can block an important project. Utility managers should be transparent and highlight the range of project benefits that affect the community. Activities can range from passive outreach, such as bill enclosures and Web pages, to highly interactive outreach such as community presentations and developing school curriculum. For guidance, the reader is referred to *Survival Guide for Public Communications* (WEF, 2002) and other resources included in Section 10. Developing aesthetic outreach material can help rebrand wastewater operations to highlight sustainability activities. Utilities of all sizes and budgets have successfully created print and online material to support communications plans. DC Water in Washington, D.C., chose to use printed materials (see Figure 4.1), while utility managers in Brattleboro, Vermont, created a 50-minute online video explaining the need for a water resource recovery facility (WRRF) upgrade. Small demonstration projects such as the 1.8-kW wind turbine at the Saco,

---

Community benefits of energy projects.

☑ Local jobs

☑ Long-term rate benefit

☑ Availability of external funds to support energy generation and efficiency projects

☑ New commercial opportunities

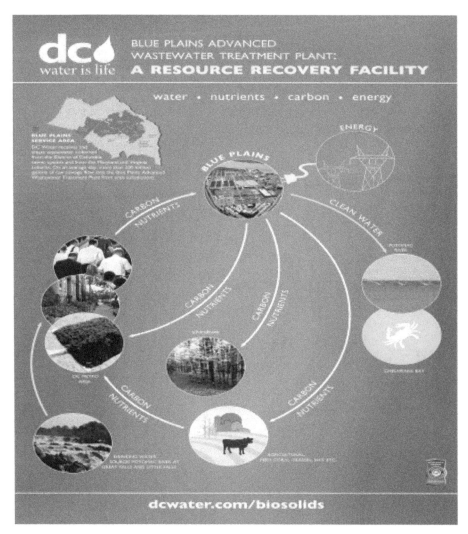

FIGURE 4.1   Example of DC Water customer outreach material.

Maine, WRRF can be useful for creating dialogue about energy even if the actual energy effects are small.

Regulatory and Legislative Agencies

Regulatory agency staff can be strong allies or staunch opponents. Therefore, an outreach program should focus on establishing a high level of trust with utility regulators. The first step is to develop good working relationships with key individuals at the appropriate regulatory agencies. As a utility expands its purview to include resource recovery and/or energy generation, utility managers are likely to face more situations in which project approvals involve multiple

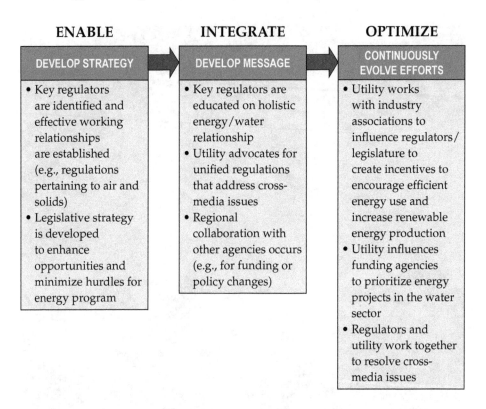

ENABLE	INTEGRATE	OPTIMIZE
DEVELOP STRATEGY	DEVELOP MESSAGE	CONTINUOUSLY EVOLVE EFFORTS
• Key regulators are identified and effective working relationships are established (e.g., regulations pertaining to air and solids) • Legislative strategy is developed to enhance opportunities and minimize hurdles for energy program	• Key regulators are educated on holistic energy/water relationship • Utility advocates for unified regulations that address cross-media issues • Regional collaboration with other agencies occurs (e.g., for funding or policy changes)	• Utility works with industry associations to influence regulators/legislature to create incentives to encourage efficient energy use and increase renewable energy production • Utility influences funding agencies to prioritize energy projects in the water sector • Regulators and utility work together to resolve cross-media issues

regulatory areas. Resolving these cross-media (water, air, and solids) issues will depend on the ability of utility managers to effectively convey project benefits and advocate for a sound regulatory approach. To carry out this advocacy, collaborating with other agencies in a utility's region or even regional and statewide associations can greatly reduce the burden.

Media Outreach

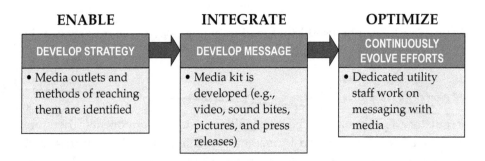

ENABLE	INTEGRATE	OPTIMIZE
DEVELOP STRATEGY	DEVELOP MESSAGE	CONTINUOUSLY EVOLVE EFFORTS
• Media outlets and methods of reaching them are identified	• Media kit is developed (e.g., video, sound bites, pictures, and press releases)	• Dedicated utility staff work on messaging with media

Similar to regulators, the media can represent a utility's strongest ally or worst enemy. Typically, the best approach is to be proactive and create a positive story. After identifying key media outlets, utility managers should develop a media strategy and create a media kit with a video, sound bites, photographs, fact sheets, and press releases. If an organization does not already have a public relations professional, one or two members of a utility's staff should be trained in media relations. Utility managers should invite media to important events that they wish to have highlighted to the public.

When communicating with members of the media, it is important to be open and transparent. It is okay to say, "I do not know the answer, but I will find out and get back to you". Indeed, coming up with a hasty response on the spot can often lead to misinformation being conveyed.

> **Media communication tips**
>
> ☑ Be responsive (media tend to work on tight deadlines)
>
> ☑ Be open and transparent (it is okay to say, "I don't know the answer and will get back to you")
>
> ☑ Stay on message (don't get pulled into an area that you are not prepared to discuss)

It is also important for utility managers to stay on message and to not get pulled into discussing an area that they are not prepared to address. Managers will have the best luck getting an accurate story if they are able to explain the program or project clearly in jargon-free language, avoiding the use of acronyms. Using visuals (e.g., sketches, drawings, and photographs) can help a targeted lay audience better understand technical or scientific information. Finally, when contacted by the media, it is important to be responsive. Utility managers should keep in mind that media representatives, especially broadcast media, are likely working on a tight schedule with short turnaround times.

Environmental Advocacy Groups

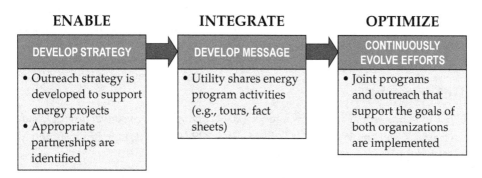

ENABLE	INTEGRATE	OPTIMIZE
DEVELOP STRATEGY	**DEVELOP MESSAGE**	**CONTINUOUSLY EVOLVE EFFORTS**
• Outreach strategy is developed to support energy projects • Appropriate partnerships are identified	• Utility shares energy program activities (e.g., tours, fact sheets)	• Joint programs and outreach that support the goals of both organizations are implemented

Environmental advocacy groups often have an excellent conduit to the media and can assist in gaining more widespread attention for an energy program. In addition, they can be excellent partners when there is a shared goal. Indeed, working together on a utility's energy program, in which goals are aligned, may make it easier to work together on other activities in which objectives may differ.

Multiple benefits of energy projects.

☑ Energy security and sustainability

☑ Reduces air pollution and greenhouse gas emissions

☑ Reduces water effects of energy production (e.g., oil spills, "fracking", and coal ash storage)

☑ Local jobs

Water Industry

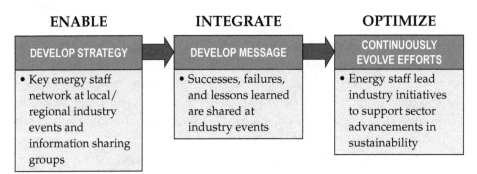

ENABLE	INTEGRATE	OPTIMIZE
DEVELOP STRATEGY	**DEVELOP MESSAGE**	**CONTINUOUSLY EVOLVE EFFORTS**
• Key energy staff network at local/ regional industry events and information sharing groups	• Successes, failures, and lessons learned are shared at industry events	• Energy staff lead industry initiatives to support sector advancements in sustainability

Communicating within the water treatment industry provides many benefits, including enhanced understanding of industry practices and state-of-the-art technologies. The level of effort can range from a one-time Web cast to a local organization evening meeting, one-day seminars, or a multiple-day state or national conference attended by thousands. Sharing information on successes, failures, and lessons learned in energy management will help move the water treatment industry together toward greater sustainability.

References

U.S. Environmental Protection Agency (2012) *Planning for Sustainability: A Handbook for Water and Wastewater Utilities*; EPA-832/R-12-001; U.S. Environmental Protection Agency: Washington, D.C.

Water Environment Federation (2002) *Survival Guide for Public Communications*; Water Environment Federation: Alexandria, Virginia.

Section 5

Demand-Side Management

"Because saving electricity means that the utility saves not only the energy it uses but also the additional energy consumed during generation, an energy conservation and peak demand reduction effort saves money, reduces carbon footprint, and jumpstarts the facility on its path toward energy sustainability".

David J. Reardon, Senior Vice President, HDR Engineering,
Folsom, California

The demand-side management topic area comprises the following themes:

- Electricity costs and billing,
- Electric power measurement and control,
- Energy management, and
- Source control.

In this document, *demand-side management* refers to programs that target reduction of both peak power demand (kilowatts or kilovolt amperes) and energy consumption (kilowatt-hours) by implementing better load and energy management strategies. Demand-side management programs promote energy efficiency and aim to reduce demand for energy (electricity or gas) and/or to shift energy demand from peak to off-peak times and possibly improve power factor (a measure of how efficiently consumed power is used). The discussion that follows is primarily focused on electricity, which typically represents the most intensive energy consumption at a WRRF. The approach to managing use of natural gas and other fuels is similar in concept to managing electricity.

Electricity Costs and Billing

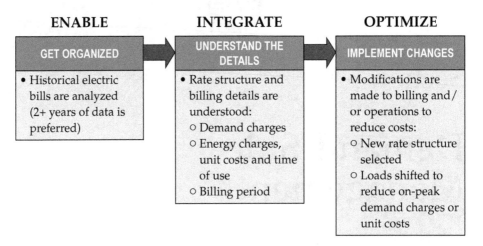

ENABLE	INTEGRATE	OPTIMIZE
GET ORGANIZED	**UNDERSTAND THE DETAILS**	**IMPLEMENT CHANGES**
• Historical electric bills are analyzed (2+ years of data is preferred)	• Rate structure and billing details are understood: ○ Demand charges ○ Energy charges, unit costs and time of use ○ Billing period	• Modifications are made to billing and/or operations to reduce costs: ○ New rate structure selected ○ Loads shifted to reduce on-peak demand charges or unit costs

It is important that staff at every level understand how a utility uses and is charged for electricity. Real-time Web access to the electricity account may be possible and is encouraged. Development of databases of energy, demand, and flow are important management metrics. Figure 5.1 provides an example of a typical electric bill, with important components highlighted.

Rate Schedule and Billing Period

The rate schedule for a utility should be noted on its bill. The electric utility typically determines the utility's rate schedule based on the type and size of the facility; however, a utility may not be under the correct rate schedule or may have other options. For example, if loads are able to be curtailed (or if the utility is able to pay a premium) during peak demand periods, the utility may be eligible for a lower unit cost during all other times of use. It is important to remember that load shedding using standby power is discouraged by many air regulation agencies. In addition, the billing period is typically 1 month but does not necessarily start on the first day of the month. Knowing when the billing period starts and ends will be important for reducing a utility's monthly peak demand charge.

Energy Charges and Time of Use

The energy charge (dollars/kilowatt-hour) can be billed at a set rate throughout a billing period or it can be based on time of use, with maximum rates charged during the electric utility peak demand hours (typically for 4 or more hours during weekday afternoons). The variability between peak and off-peak hours can be significant; in the example bill shown in Figure 5.1, the peak power unit cost is close to double the off-peak power unit cost.

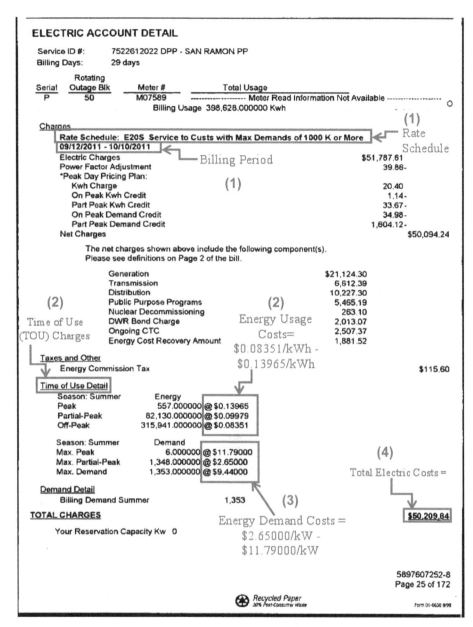

FIGURE 5.1 Example of an electric bill and key components.

Demand Charges

Demand is the maximum power drawn, typically expressed as kilowatts. The demand charge (dollars/kilowatt) is assessed for the maximum demand during the billing period. It is measured over a specific, contiguous period of time, typically 15 or 30 minutes. It represents the combined power draw for all electrical devices. This charge can be significantly affected by short-duration operation of a large motor such as an aeration blower, pump, or centrifuge. Cost savings can be realized through the control of maximum demand during a billing period. Although this strategy does not save energy use directly, it is part of an overall strategy for controlling energy costs. Sometimes, as shown in the example bill, the demand charge varies by time of use. Costs can be reduced by avoiding starting up certain types of equipment with a high power draw during peak hours.

Total Electricity Costs

Total electricity costs include electricity use charges (the sum of energy or generation, transmission, and distribution), demand charge(s), and taxes and other miscellaneous charges. Calculating the relative proportion of electricity use and demand charge can be used to determine which area to focus on first.

Electric Power Measurement and Control

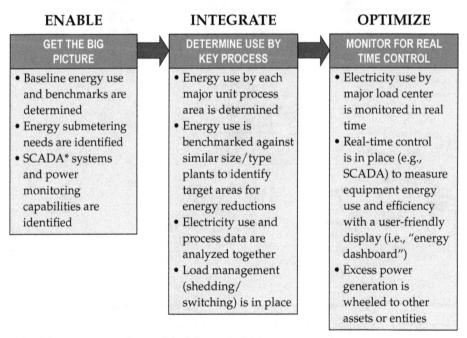

ENABLE	INTEGRATE	OPTIMIZE
GET THE BIG PICTURE	**DETERMINE USE BY KEY PROCESS**	**MONITOR FOR REAL TIME CONTROL**
• Baseline energy use and benchmarks are determined • Energy submetering needs are identified • SCADA* systems and power monitoring capabilities are identified	• Energy use by each major unit process area is determined • Energy use is benchmarked against similar size/type plants to identify target areas for energy reductions • Electricity use and process data are analyzed together • Load management (shedding/switching) is in place	• Electricity use by major load center is monitored in real time • Real-time control is in place (e.g., SCADA) to measure equipment energy use and efficiency with a user-friendly display (i.e., "energy dashboard") • Excess power generation is wheeled to other assets or entities

*SCADA = supervisory control and data acquisition.

Benchmarking—The Big Picture

Benchmarking of energy intensity (kilowatt-hour/flow volume) against other plants with similar processes and treatment goals can be useful to begin the energy optimization process. Utility managers should compare the energy intensity of their utilities to data provided in the literature. One such study with useful data was conducted by the Electric Power Research Institute (Goldstein and Smith, 2002). The U.S. EPA also provides unrestricted access to a benchmarking tool tailored specifically for water resource recovery facilities as part of the ENERGY STAR Program (U.S. EPA, 2012). Guidance on benchmarking is also provided in U.S. EPA's publication, *Ensuring a Sustainable Future: An Energy Management Guidebook for Wastewater and Water Utilities* (U.S. EPA, 2008). Another useful tool is the Water Environment Research Foundation's CHEApet (WERF, 2011).

Benchmarking—The Details

If submetering data are available, examining energy use by process area will help to determine which areas deserve the most focus. In addition, data may be compared to benchmarks of typical facilities to reveal areas in which a utility's use is significantly higher, indicating potential opportunities for improvement (e.g., see the pie chart of typical energy use by process area shown in Figure 5.2). During benchmarking activities, utility managers should identify locations to consider installing submetering to facilitate better data collection in the future.

Process Data

Analyzing energy use compared with process data provides a better understanding of how process conditions and operating choices affect energy use. A simple plot of daily flow data and energy use can reveal inefficiencies. Energy

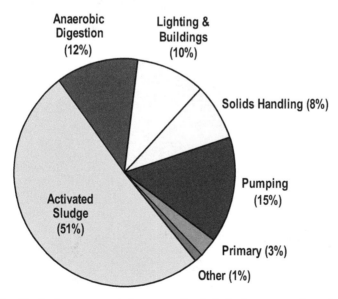

FIGURE 5.2 Typical energy use for an activated sludge secondary facility (SAIC, 2006; WEF, 2009).

use versus influent conditions (e.g., biochemical oxygen demand) also provides a benchmark that can be compared to literature values. Assessing a variety of operating conditions compared with energy use can help determine key parameters for future monitoring. This concept can be advanced further by developing a process model that integrates energy and process data.

Monitoring

Successful energy management incorporates real-time monitoring and control. Various metrics can be measured, such as pump efficiency or the energy intensity of a process or the entire facility (kilowatt-hour/volume treated is a common metric). Software programs can integrate these concepts to supervisory control and data acquisition (SCADA) systems to provide instant feedback to operators for real-time decision making. Figure 5.3 presents an example of an "energy dashboard" that provides a snapshot of key energy monitoring data.

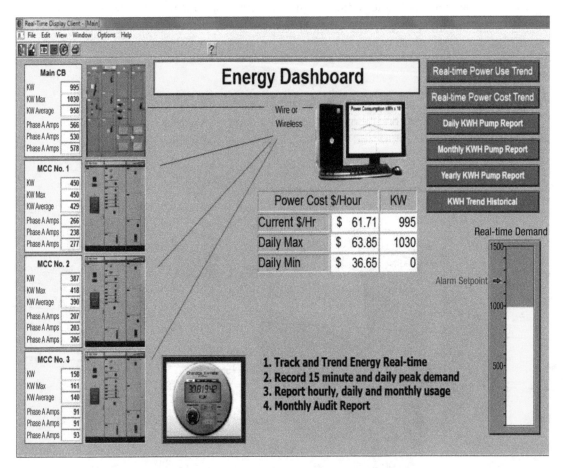

FIGURE 5.3   Sample energy dashboard screen shot. (Intellisys, Inc. Energy Management Software www.intellisyssoftware.com <http://www.intellisyssoftware.com>)

Energy Management

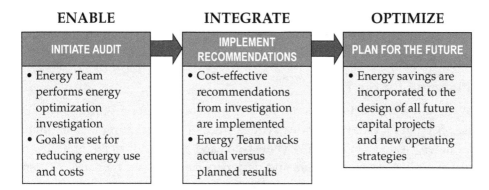

ENABLE	INTEGRATE	OPTIMIZE
INITIATE AUDIT	**IMPLEMENT RECOMMENDATIONS**	**PLAN FOR THE FUTURE**
• Energy Team performs energy optimization investigation • Goals are set for reducing energy use and costs	• Cost-effective recommendations from investigation are implemented • Energy Team tracks actual versus planned results	• Energy savings are incorporated to the design of all future capital projects and new operating strategies

Conducting an Energy Audit or Energy Optimization Investigation

The heart of demand-side management is the energy optimization process in which a systematic methodology is used to reduce energy use to the lowest practical level. For existing facilities, this task is often called an *energy audit* (ASHRAE, 2011). Unfortunately, this term has negative connotations and, as such, the process is probably better described as *energy optimization*. Regardless of the title, this task focuses on facility unit operations, processes, and equipment so that, in aggregate, the facility can achieve superior energy-intensity benchmarking values compared to facilities of its type and size. An energy optimization study could be conducted using a top-down approach

Reduce energy use of key equipment by

☑ Shutting down

☑ Operating part time

☑ Operating with variable speed

☑ Operating at lower flows

☑ Operating at lower pressures

☑ Upgrading with more efficient equipment

1. Is the equipment or process oversized or improperly sized for current conditions? Oversized or improperly sized equipment is a significant cause of excessive energy consumption.

2. Are process setpoints reasonable (e.g., return activated sludge flows, activated sludge dissolved oxygen levels, wet well levels)?

3. Would using variable-speed control allow equipment to operate at desired conditions most efficiently?

4. Can equipment flow or pressure be reduced?

5. Can equipment be replaced with smaller or more efficient models?

6. Can equipment be operated part time (e.g., intermittent mixing or channel air)?

7. Is the equipment or process operating near maximum efficiency or are improvements required?

8. Can the process function be accomplished in a different way with lower energy use (e.g., alternative activated sludge process configurations, fewer tanks, lower solids residence time)?

9. Does equipment require maintenance to improve performance (e.g., worn pump impellers and damaged wear rings)?

10. Can equipment be shut down? Some equipment can be shut down with little or no effect on facility performance.

based on facility benchmark comparisons or a bottom-up approach based on individual pieces of equipment. The benchmarking approach is discussed briefly in "Benchmarking—The Details" section. The equipment survey approach identifies all significant processes and pieces of equipment in the facility to determine whether they could be operated differently to reduce energy use.

Planning for the Future

While the water sector may aim for energy neutrality for all future projects (see Section 7), today the industry must focus on configuring conventional treatment systems to minimize energy demand and optimize the production of renewable energy. Planners and designers can consider the latest proven technologies to reduce energy use. Table 5.1 presents a list of approaches to minimize energy demand (and generate energy) through planning and design of new facilities.

Source Control

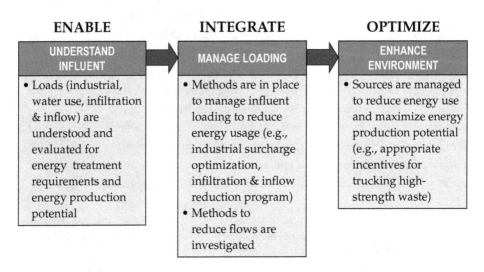

ENABLE	INTEGRATE	OPTIMIZE
UNDERSTAND INFLUENT	**MANAGE LOADING**	**ENHANCE ENVIRONMENT**
• Loads (industrial, water use, infiltration & inflow) are understood and evaluated for energy treatment requirements and energy production potential	• Methods are in place to manage influent loading to reduce energy usage (e.g., industrial surcharge optimization, infiltration & inflow reduction program) • Methods to reduce flows are investigated	• Sources are managed to reduce energy use and maximize energy production potential (e.g., appropriate incentives for trucking high-strength waste)

Understanding both flow and influent loading will help utility managers evaluate process performance and baseline energy use. If flows or loading change, treatment and energy optimization will be affected. Different strategies are required for different materials. For example, reducing flows by addressing infiltration and inflow should be encouraged because it will reduce energy needs and costs. When using green infrastructure to encourage groundwater infiltration, a utility may reduce its wet weather flow and obtain additional benefits through the use of green infrastructure practices. Historically, high-strength surcharge programs have encouraged industrial users to reduce organic loading to the system. However, when using anaerobic digestion, a utility may want that

TABLE 5.1   Approaches to minimize energy use of future facilities.

Opportunity	Approach
Optimize primary clarifier performance to efficiently divert organics before aerobic systems, thereby decreasing energy demand and decreasing biological waste production. This increases load to anaerobic digesters and increases energy generation.	• Optimize primary clarifier design and hydraulic loading. • Add chemicals to enhance primary clarification. • Use dissolved air flotation. • Treat raw wastewater or settled primary effluent with short contact process activated sludge to enhance clarification (e.g., Strass, Austria, imZillertal Wastewater Treatment Plant [WERF, 2009]).
Minimize aeration system energy demand.	• Specify/install aeration systems with energy-efficient blowers (e.g., single-stage blowers or air-bearing blowers). • Use fine-bubble diffusers. • Consider emerging technologies such as ultra-high-efficiency strip aeration. • Install automated dissolved oxygen control. • Consider swing zones that can be used for either anoxic or aerobic treatment, depending on loading conditions.
Use equipment that varies output.	• Use variable-frequency drives for efficient treatment throughout diurnal flow and loading rate variation.
Properly size of equipment and processes.	• Consider shorter planning horizons and more frequent capacity expansion to ensure that equipment and processes are used in an energy-efficient manner. • Evaluate existing plant flow and load conditions compared to design conditions.
Optimize plant hydraulic profile.	• Develop a hydraulic design so that pumping is minimized and the hydraulic profile is optimized.
Use plant flow equalization or system peak shaving to minimize flow/load variation.	• Consider collection system equalization via smart grid technology or larger systems such as deep tunnel storage to minimize peaking events that lead to oversizing of equipment/systems.
Use automation/SCADA/computer technology to minimize energy.	• Use instruments and automation appropriately to optimize system efficiency. • Consider having an energy dashboard integrated to the SCADA system to monitor and record energy metrics. • Use real-time measurement of demand to optimize processes and minimize energy use.
Consider how auxiliary processes affect overall facility energy demand. Evaluate the benefit of these processes vs the environmental effect from increased energy demand.	• Evaluate how to reduce energy use in odor control, water reuse, tertiary treatment (to meet low effluent limits), and disinfection systems.

extra loading to enhance biogas production. Therefore, setting up source control programs to divert high-strength waste from sewers and instead introducing it directly for digestion (i.e., through trucked waste programs) may be the most cost-effective strategy.

References

American Society of Heating, Refrigeration, and Air-Conditioning Engineers (2011) *Procedures for Commercial Building Energy Audits*, 2nd ed.; American Society of Heating, Refrigeration, and Air-Conditioning Engineers: Atlanta, Georgia.

Goldstein, R.; Smith, W. (2002) *Water & Sustainability (Volume 4): U.S. Electricity Consumption for the Next Half Century*; Electric Power Research Institute: Palo Alto, California.

Science Applications International Corporation (2006) *Water and Wastewater Energy Best Practice Guidebook*; Developed for Focus on Energy, Madison, Wisconsin.

U.S. Environmental Protection Agency (2008) *Ensuring a Sustainable Future: An Energy Management Guidebook for Wastewater and Water Utilities*; Office of Wastewater Management; U.S. Environmental Protection Agency: Washington, D.C.

U.S. Environmental Protection Agency (2012) Energy Star Portfolio Manager. http://www.energystar.gov/index.cfm?c=evaluate_performance.bus_portfoliomanager (accessed Jan 2013).

Water Environment Federation (2009) *Energy Conservation in Water and Wastewater Treatment Facilities*; WEF Manual of Practice No. 32; McGraw-Hill: New York.

Water Environment Research Foundation (2009) *Best Practices for Sustainable Wastewater Treatment: Initial Case Study Incorporating European Experience and Evaluation Tool Concept*; Project No. OWSO4R07a; Water Environment Research Foundation: Alexandria, Virginia.

Water Environment Research Foundation (2011) *Demonstration of the Carbon Heat Energy Assessment and Plant Evaluation Tool (CHEApet)*; Project No. OWS-O4R07g; Water Environment Research Foundation: Alexandria, Virginia.

Section 6

Energy Generation

"The wastewater sector is beginning the transition from wastewater treatment to resource recovery, including becoming a source of distributed energy generation".

Ralph ("Rusty") B. Schroedel, Jr., Executive Engineer,
Brown and Caldwell, Milwaukee, Wisconsin

The energy generation topic area comprises the following themes:

- Strategy,
- Energy from water,
- Supplemental energy sources, and
- Renewable energy certificates (RECs).

Energy Generation Strategy

ENABLE	INTEGRATE	OPTIMIZE
SET PRODUCTION GOAL	**OBTAIN SUPPORT**	**GROW PROGRAM**
• Measurable energy-generation goal is established • Energy-generation plan is coordinated with utility strategic plan • Energy Team understands regulatory and permit limitations (e.g., air emissions) with regard to generation	• Governing body approves capital budget for energy-generation projects • Regulatory issues have been addressed and satisfactorily resolved	• Infrastructure for energy generation is proactively maintained, renewed, and upgraded • Holistic evaluation methodologies (e.g., triple bottom line) are used to evaluate energy-generation opportunities

Establishing Energy Goals

Whether a utility has been generating on-site energy for decades or is considering it for the first time, a first step is to establish measurable energy goals that need to be embraced at all levels of the organization. The goal could be as simple as generating a portion of electricity needs using on-site renewable sources or more aggressive, such as meeting all on-site electricity demands using on-site renewable sources by the year 2020. Integrating the goal to the strategic planning process (as discussed in Section 3) helps ensure that it will be carried forward and incorporated to capital budgeting.

Planning for Implementation Obstacles

To establish an energy generation program, staff must have a strong understanding of the potential regulatory obstacles (e.g., National Pollutant Discharge Elimination System and air permits), technical requirements (e.g., operator and maintenance skills and training), technical constraints (e.g., digester gas conditioning), and transactional complexities (e.g., identifying and contracting with customers for electricity or fuel produced). It is important to learn from other utilities that have been generating energy on site in a utility's region (or state) to help identify key considerations.

Identifying Energy Sources

A utility's energy-generation strategy may include some combination of inherent energy in water and supplemental energy sources. In drinking water operations, the most common form of inherent energy is kinetic energy that can be harvested as source water flows from higher to lower elevations (i.e., hydropower). In wastewater operations, the most common form of inherent energy comes from anaerobic digestion, whereby digester gas is converted to electricity, fuel, or hot water using boilers, engines, or gas turbines. Supplemental energy sources most commonly include wind and solar power or, for wastewater, trucked organic wastes to augment digester gas production ("co-digestion").

Energy from Water

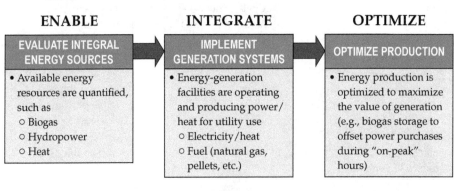

Biogas

Digester gas, or biogas, is a byproduct of anaerobic digestion, which is used to stabilize biosolids generated during wastewater treatment. Typical biogas from digestion of municipal biosolids contains 60 to 65% methane (natural gas). As Table 6.1 shows, numerous water resource recovery facilities of all sizes worldwide have been using biogas for energy production for at least the last 20 years and many are planning new projects today.

Use of digester gas for electricity, combined with the beneficial use of the waste heat, is referred to as *combined heat and power* (CHP). Historically, biogas has been used mostly in these applications. More recently, biogas has been used for transportation fuel or directly used or sold as renewable natural gas ("biomethane"). A utility's preferred choice of product and technology depends on many factors, including site conditions, regulatory requirements, quantity of biogas available, current value of electrical power, fuel and RECs, and on-site energy needs (see Figure ES-2).

Digester gas may be used to produce electricity with technologies such as microturbines, internal combustion engines, gas turbines, fuel cells, steam turbines, or sterling engines. Alternatively, biogas may be used to fire boilers or dryers and run motors or be converted to a gaseous or liquid fuel for use in fleet vehicles or to sell (see Figure ES-2).

There are many resources available to help utility managers select the most appropriate biogas use technology, including

- *Life Cycle Assessment Manager for Energy Recovery (LCAMER)*—this report which were commissioned by the Water Environment Research Federation (2012c) in 2007 and updated in 2012, includes a detailed spreadsheet tool (LCAMER) to assist users in the selection of technology for energy recovery from biogas;

Biogas engineering guidelines.

- Anaerobic digestion with biogas recovery is most common at water resource recovery facilities that treat at least 20 ML/d (5 mgd) of wastewater because of economical feasibility constraints, although smaller plants have also successfully implemented anaerobic digestion (WERF, 2012a).

- The heating value of biogas produced by anaerobic digesters is approximately 24.3 mJ/m³ (600 Btu/cu ft) (Metcalf and Eddy, 1991).

- Approximately 0.9 m² (1 sq ft) of digester gas can be produced by an anaerobic digester per person per day (Metcalf and Eddy, 1991). This volume of gas can provide approximately 2.2 W of power.

TABLE 6.1 Examples of facilities using biogas for energy generation.

Location	Average flow (ML/d)	Rate ($/kWh)	Type (year installed)	Total capacity (kW)	Capital cost ($)	Annual net savings ($/yr)	Comments
Albert Lea, Minnesota	16 (4.3 mgd)	$0.07	Microturbines (2003)	120	$92,000	$70,000	Power utility paid for most of original installation costs and operation and maintenance for 5 years. Sold to Albert Lea for $1.
Burlingame, California	13 (3.4 mgd)	$0.14	Internal combustion engine (2006)	200	$912,000	$116,000	
Marshalltown, Iowa	24 (6.4 mgd)	$0.07	Internal combustion engine (1972)	1000	Unknown	$70,000	
Sheboygan, Wisconsin	40 (10.5 mgd)	$0.06	Microturbines (2006, 2010)	700	$1.8 million	$125,000–250,000	Power utility paid for most of 2006 purchase, later sold to Sheboygan.
Gwinnett County/ Fort Wayne Hill, Georgia	113 (30 mgd)	$0.04	Internal combustion engine (2011)	2100	$5.2 million	$1 million	
Narrangansett Bay/ Bucklin Point, Rhode Island	83 (22 mgd)	$0.11	Internal combustion engine (2014)	600	$4-5 million	$325,000	Planned
Occoquan, Virginia	(32 mgd)	$0.06	Internal combustion engine (2013)	848	4.5 million	$375,000	Planned
Hampton Roads, Virginia	204 (54 mgd)	$0.06	Internal combustion engine (2012–2013)	2200	$406,000/yr, 20-year debt service	$385,000	

56

- *Evaluation of Combined Heat and Power Technologies for Wastewater Treatment Facilities*—this report was commissioned by U.S. EPA in 2010 and includes an overview and comparison of biogas utilization technologies, including cost ranges (Wiser et al., 2012);

- *Opportunities for Combined Heat and Power at Wastewater Treatment Facilities: Market Analysis and Lessons from the Field*—this report was commissioned by U.S. EPA in 2011 to update the initial report from 2007 and includes an assessment of the market potential for CHP at U.S. WRRFs;

- U.S. EPA Combined Heat and Power Web site (http://www.epa.gov/chp/)—this Web site contains additional information on CHP applications and tools for project development; and

- Water Environment Federation Biogas Data Web site (http://www.biogasdata.org)—this Web site, which was launched in 2012, contains basic data on biogas use in the United States.

Hydropower

Large hydroelectric projects can generate a significant amount of energy as source water is conveyed from higher to lower elevations. In drinking water and wastewater operations, there are new opportunities to harness the hydrokinetic energy of flowing water and wastewater for electricity production. These options include in-conduit and microhydropower turbines. The energy produced is a function of the volume of water released (discharge) and the vertical distance the water falls (head).

Hydropower in the City of Dover, New Hampshire.

The City of Dover, New Hampshire, demonstrated the use of an innovative small hydropower technology that uses an 18-m (59-ft) vertical drop in its outfall pipe. The WRRF discharges approximately 11 ML/d (2.9 mgd) and expects to produce approximately 15 kW.

Heat Pumps

A heat pump is a device that uses a small amount of energy to move heat from one location to another. The wastewater or effluent at a WRRF can be used as a heat source or sink. A heat pump can be used to extract and transfer heat to provide heating or cooling energy for facility buildings or process uses. Generally, economics are more favorable in colder climates and locations where fossil fuel prices are high. There are various examples of facilities in North America that recover heat from their effluent, including Chicago, Illinois; Kent County, Delaware; King County, Washington; and Philadelphia, Pennsylvania.

Other Inherent Energy Sources

Chemical energy in the liquid stream of wastewater, which is typically treated aerobically and requires an energy input, could theoretically be recovered to produce energy if a different treatment process were used (e.g., anaerobic treatment or microbial fuel cell). Similarly, treated biosolids have inherent energy that currently is not typically captured because of the energy requirements to dry biosolids (see Alternative Treatment Technologies in Section 7 for a more detailed discussion of emerging technologies).

Supplemental Energy Sources

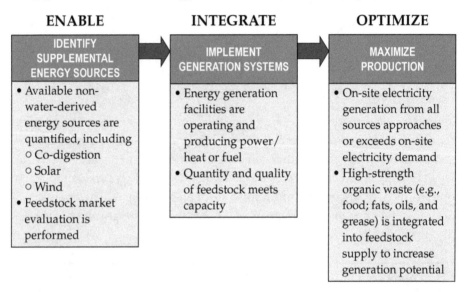

Supplemental energy sources include any energy source used that is not directly derived from water. The three most common types of supplemental sources (co-digestion, solar power, and wind power) are covered in this section. The Water Environment Research Foundation developed *Green Energy Life Cycle Assessment Tool (GELCAT) and User Manual* to evaluate the economic viability of renewable energy technologies (WERF, 2012b). Another potential energy source is geothermal energy taken from the ground or from groundwater.

Co-Digestion

During the past decade, a number of utilities have expanded their digester operations to include additional high-strength organics such as food processing waste; fats, oils, and grease (FOG); and, more recently, food scraps. These materials are typically rapidly degradable and can significantly increase biogas production. In addition, providing a sustainable recovery option reduces the amount of material that is illegally disposed or deposited in a landfill. Anaerobic digestion of material that would otherwise be landfilled reduces GHG emissions because all methane produced in the anaerobic digesters can be captured and converted to carbon dioxide (a much less potent GHG). In 2010, WEF published a technical practice update devoted to the subject of high-strength waste co-digestion in municipal anaerobic digesters (WEF, 2010).

Solar Power

Solar power, or photovoltaic energy, has become increasingly common and more cost-effective. The average factory price of a photovoltaic panel is approximately $1/W, which results in an installed cost of $2500 to $5000/kW with auxiliary equipment, support structures, installation, and insurance (U.S. EPA Combined Heat and Power Web site, http://www.epa.gov/chp/). Solar energy prices have declined an average of 4% each year over the past 15 to 20 years and are expected to continue to decrease at an accelerated rate. (For reference, in the early 1980s, photovoltaic systems cost more than $25,000/kW.)

Photovoltaic energy is modular in nature and can be situated in various areas, including on building roofs, integrated to structures, or as freestanding arrays on open land. Solar energy produces the most energy during peak hours and has low operational and maintenance requirements. Additional information can be found on the National Renewable Energy Laboratory Web site (http://www.nrel.gov/analysis/pubs_solar.html), which includes tools for estimating potential power generation for specific sites and design guidelines for system operations. Solar energy can also be used to heat water and, in some systems, improve sludge drying.

Wind Power

The ideal location for a wind turbine is a site with no obstructions to cause disruption to wind flow and one that is away from human habitation. However, in reality, such a site is rarely possible. In practice, the most important factor to consider is wind speed. Typically, a consistent wind speed of at least 19 km/h (12 mph) is required. With power being proportional to the wind speed cubed, a small increase in wind speed can have a significant effect on the power output. The wind speed can be maximized by using higher towers and siting the turbine away from obstructions. Sufficient wind energy is most prevalent in hilly terrains and along the shores of large bodies of water such as bays, lakes, and the ocean.

Wind turbine system pricing on a kilowatt basis is similar to that for photovoltaic energy. However, while photovoltaic energy only operates at an annual average of 6 hours per day, wind may operate up to 24 hours per day, producing more energy per dollar invested. On the other hand, wind turbines have significantly more operation and maintenance requirements than photovoltaic energy.

Renewable Energy Certificates

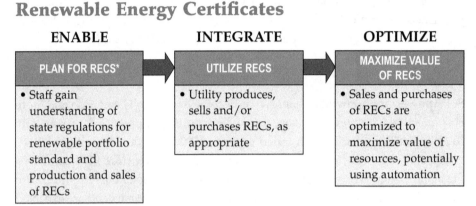

*REC = renewable energy certificate.

Renewable energy certificates represent the monetized, environmental attributes of power generated from renewable sources. Renewable energy certificates can be bought by electric utilities or generators to comply with renewable energy mandates. In states with requirements for renewable energy production, RECs can be used for compliance. In states without these requirements, RECs are traded in the voluntary market.

Renewable Portfolio Standard

Renewable portfolio standard (RPS) policies mandate utilities to own or acquire renewable energy or RECs to account for a certain percentage of their total electricity sales or a certain amount of generating capacity within a specified timeframe. As discussed in Toward Carbon Neutrality in Section 2, many U.S. states, Canadian provinces, and countries worldwide have enacted RPS or similar policies or goals. Several states with RPS policies, such as California, Colorado, Florida, Massachusetts, and New York, consider wastewater biogas and biomass fuels to be eligible renewable energy resources; electricity generated using wastewater biogas and residuals contribute to RPS goals or requirements. Some states also provide grants and financial incentives to develop and encourage the use of renewable fuels and renewable electricity generation. More information on state RPS policies and goals can be obtained from the Database of State Incentives for Renewables & Efficiency (http://www.dsireusa.org).

Buying and Selling Renewable Energy Certificates

Renewable energy certificates are typically purchased through REC brokers who enter into forward-pricing agreements with owners of the renewable assets. As the market matures, utility managers can expect sophisticated REC trading automation to come online similar to what currently exists for electricity or other commodities. Renewable energy certificates are priced on a dollar per megawatt-hour basis. The primary purchasers of RECs include

- Utilities to meet their RPS goals,
- Large corporations to meet their sustainability goals, and
- Speculators who believe that future drivers (e.g., carbon tax or cap and trade program) will increase REC values.

As regulations of GHG emissions become more widespread, the value of RECs will increase. In states with RPS policies already in place, the value of RECs can be significant.

Example of the value of RECs.

In 2012, East Bay Municipal Utility District entered into a power purchase agreement to sell excess renewable energy generated on site. The agreement resulted in a unit price double what was available on the wholesale market by including a payment for the REC at approximately the same value as that of the energy.

Green Power Purchasing Policies

Many local governments are committed to using a certain percentage of their power from renewable sources, including wastewater biogas and biomass fuels, thus mandating the purchase of green power. There are opportunities for utilities to enhance their green energy use by purchasing green power directly from power utilities or by purchasing RECs. In addition, if a utility has on-site renewable energy generation, the utility may have the option to sell excess power or the RECs associated with it to local governments that need to comply with green power purchasing policies. Another option is to invest in green power facilities. More information on green power purchasing policies, goals, and eligible technologies can be obtained at Open Energy Info's Web site (http://en.openei.org/wiki/Green_Power_Purchasing).

Another similar effort to increase renewable energy in the marketplace is community choice aggregation (CCA), whereby local governments may purchase electricity for the electric users within their service area, while retaining the transmission and distribution services of the local utility. The CCA typically offers a much higher percentage of renewable energy than is provided by the investor-owned utility (as high as 100%). Six U.S. states have enacted CCA laws (i.e., California, Illinois, Massachusetts, New Jersey, Ohio, and Rhode Island) and fewer than a dozen programs are currently operating (Department of Energy: http://www.dsireusa.org and http://apps3.eere.energy.gov/greenpower/markets/community_choice.shtml).

References

Metcalf and Eddy (1991) *Wastewater Engineering: Treatment, Disposal, Reuse,* 3rd ed.; McGraw-Hill: New York.

U.S. Environmental Protection Agency (2011) *Opportunities for Combined Heat and Power at Wastewater Treatment Facilities: Market Analysis and Lessons from the Field*; U.S. Environmental Protection Agency: Washington, D.C.

Water Environment Federation (2010) *Direct Addition of High-Strength Organic Waste to Municipal Wastewater Anaerobic Digesters*; Technical Practice Update; Water Environment Federation: Alexandria, Virginia.

Water Environment Research Foundation (2012a) *Barriers to Biogas Use for Renewable Energy*; Project No. OWSO11c10; Water Environment Research Foundation: Alexandria, Virginia.

Water Environment Research Foundation (2012b) *Green Energy Life Cycle Assessment Tool (GELCAT) and User Manual*; Project No. OWSO6R07c; Water Environment Research Foundation: Alexandria, Virginia.

Water Environment Research Foundation (2012c) *Life Cycle Assessment Manager for Energy Recovery (LCAMER)*; Water Environment Research Foundation: Alexandria, Virginia.

Wiser, J.; Schettler, J.; Willis, J. (2012) *Evaluation of Combined Heat and Power Technologies for Wastewater Treatment Facilities*; EPA-832/R-10-006; Prepared for Columbus Water Works, Columbus, Georgia; Brown and Caldwell: Atlanta, Georgia.

Section 7

Innovating for the Future

"A strong commitment to innovation today will attract the human talent needed to solve world challenges tomorrow".

Bill Toffey, Sustainability Strategist, Effluential Synergies, LLC,
Philadelphia, Pennsylvania

Innovating for the future comprises the following themes:

- Research and development,
- Risk management,
- Alternative treatment technologies, and
- Alternative management approaches.

Drivers for innovation include more stringent regulatory standards, tighter budgets, challenging community sustainability goals, and the promise of cost-saving technologies. However, many potential roadblocks exist. The water sector is inherently risk-averse and can be reluctant to make changes, even those that use well-established technologies or new technologies in a manner that present little risk to the utility. At the same time, the water sector recognizes the need to continue to support research and development of new technologies and strategies that will allow the industry to manage stricter regulations, rising costs, and resource scarcity.

Research and Development

Types of Innovative Technologies

Innovation refers to new technologies, new operational strategies, and use of existing technologies in new ways. In the context of *The Energy Roadmap,* innovation promises greater energy efficiency or improved energy generation.

63

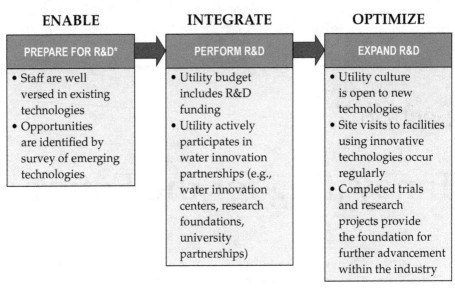

ENABLE | INTEGRATE | OPTIMIZE

PREPARE FOR R&D*	PERFORM R&D	EXPAND R&D
• Staff are well versed in existing technologies • Opportunities are identified by survey of emerging technologies	• Utility budget includes R&D funding • Utility actively participates in water innovation partnerships (e.g., water innovation centers, research foundations, university partnerships)	• Utility culture is open to new technologies • Site visits to facilities using innovative technologies occur regularly • Completed trials and research projects provide the foundation for further advancement within the industry

*R&D = research and development.

Technologies can be divided into the following three classes:

- Embryonic, which is still in the laboratory or pilot stage;

- Innovative, which is operating at a demonstration or full scale, capable of evaluation for commercial application but with very limited deployment; and

- Established, which includes commonly applied processes and technologies from other sectors.

Technologies at each stage of development provide different risks and benefits for a utility. Various resources are available to learn about embryonic, innovative, and established technologies. The Water Environment Federation's manuals of practice provide the foundation for established technologies. In addition, the Water Environment Research Foundation (WERF) produces a wealth of information on existing and emerging technologies and approaches (http://www.werf.org). U.S. EPA assists with "technology transfer"; one example is its publication summarizing more than 100 emerging technologies (U.S. EPA, 2008). Academic research articles, industry publications, and professional organization information-sharing events, widely available to agency managers and staff, offer introductions to innovative and embryonic technologies.

Creating an Innovation Hub

Although "innovation hubs" are common in energy, health, and communications sectors, they are not yet common in the water sector. Innovation hubs address a gap facing technology companies and "early adopters" in terms of access to a suite of legal, technical, and financial capacities necessary to expedite change. To address this gap in the water sector, the Leaders Innovation Forum for Technology was established in 2012 by a collaboration of WERF, WEF, and several leading public agencies to provide a mechanism for access to a bundle of services that enable a utility to take a lead in testing and development of embryonic and innovative technologies.

Using Academic Collaborations

Utility managers can be proactive in encouraging collaborations with academic researchers. For some managers, the appropriate approach to innovation is to learn from the research of others. However, other managers might join with one or more partners to sponsor university research or go further to conduct tailored research and development in house. In any case, an academic team provides highly specialized expertise and a utility provides a place to test embryonic technologies. This type of collaboration may also be able to leverage national funding support from agencies like the National Science Foundation and WERF. In some instances, a utility may be able to negotiate intellectual property rights and share in profits from commercialization of specific technologies.

Facilitating Knowledge Transfer and Innovation Management

Many professional organizations are already driving innovation. Staying involved with these organizations can be one of the simplest means to stay current on the latest advances. For example, WEF's technical conferences and governance system foster knowledge transfer around innovative technologies and their application; WERF is a leading shaper of research in the water sector. Utility managers may encourage partnerships among agencies and other utilities and technology firms to fund research and pilot testing of innovative technologies that can benefit a utility directly and the water sector more widely.

There are world-class research institutions around the country, and one may be in a utility's backyard. Utility managers may be surprised at just how much these institutions want to find out about what a utility is doing and that they want to collaborate on water research projects to a utility's benefit. Investment in research may not yield immediate results, but it ensures that multiple options are objectively explored beyond pressures of privately sponsored projects to prove results.

Looking Beyond the North American Water Sector

It is important to learn from the example of leaders in other industrial sectors and other nations. Innovations introduced in Asia and in the European Union may prove applicable in the United States. Similarly, innovative approaches in less conservative industrial sectors, such as the pharmaceutical and biofuel industries, provide models for adaptability and flexibility. The Water Environment Federation attracts speakers from around the globe to its annual technical exhibition and conference, including many who focus on emerging, innovative energy conservation and production technologies.

Managing Risk Management

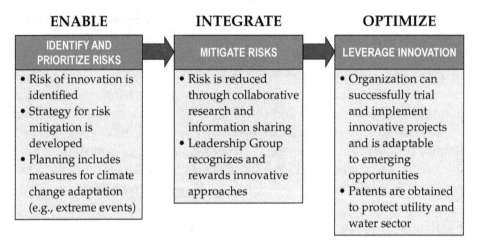

ENABLE	INTEGRATE	OPTIMIZE
IDENTIFY AND PRIORITIZE RISKS	MITIGATE RISKS	LEVERAGE INNOVATION
• Risk of innovation is identified • Strategy for risk mitigation is developed • Planning includes measures for climate change adaptation (e.g., extreme events)	• Risk is reduced through collaborative research and information sharing • Leadership Group recognizes and rewards innovative approaches	• Organization can successfully trial and implement innovative projects and is adaptable to emerging opportunities • Patents are obtained to protect utility and water sector

Developing and using new technologies carries risks, including the following:

- Technology—the water sector typically requires a technology to have multiple deployments within the same country or even region, with a long track record of successful operation. This requirement is not practicable for many innovative technologies. Agencies believe, with some justification, that new technology will require staff resources for successful deployment. They are concerned about criticism associated with a technology failure. To address these concerns, pilot testing directed by the technology company offers a means for evaluating a solution while minimizing the risks associated with full technology adoption.

- Financial—new approaches may require large capital improvements designed to operate for many decades. In addition, these improvements may best be evaluated using multidimensional tools that are less common, such as life cycle assessments and triple bottom line analyses rather than simple payback. If an innovative technology solution does not lend itself to low-cost retrofit to existing structures, a utility may need access to capital requiring utility managers to leverage incentives and use financial tools outside of their traditional area of expertise.

- Regulatory—the rigidity of regulatory permitting often provides little or no flexibility for the learning curve associated with constructing and operating new technologies. Most utility managers do not have the resources to engage with legislative and regulatory leaders to reshape permitting rules. However, they may participate in those regional and state water associations that work to seek regulations and policies in state government that accommodate innovative environmental technologies.

- Market/supply chain risk—the cost-effectiveness of energy improvements, whether efficiency or generation, is linked to future energy prices, which introduce a high degree of uncertainty in financial models. New technologies and approaches may require specific pieces of equipment that do not have a well-established market or have prices that are subject to volatility. While these risks cannot be controlled by any individual agency, the long-term picture justifies the prediction that energy prices will increase and refinements to technologies over time will improve performance toward lower energy input and costs.

Creating a Culture of Innovation

It is important to catalyze innovation by recognizing employees who have championed change. A culture of innovation applauds implementation of novel, cost-effective approaches to improve energy efficiency or generation. These can be small improvements, designed in house, or large improvements developed over time. Staff should be encouraged to present their ideas and innovations in conference sessions and sector publications. This participation can lead to increased organizational support for change and innovation.

Procurement Procedures that Minimize Risks

Contracts that promote a long view in investments in innovative technologies should be used. These can be contracts for multiple years, with flexible funding alternatives and options for multiple renewals. (The reader is referred to the discussion on alternative energy contracting in Saving Energy and Funds in Section 2). Another option for reducing financial risk is for a public agency to contract with a private partner who is a motivated technology developer to conduct a pilot or demonstration at a facility. By providing the site and, possibly, in-kind resources, a public agency can foster innovation with little upfront cost. In addition, long-term contracts for public–private partnerships can result in the co-location of significant projects for biofuel and energy generation at regional facilities, with shared risks and potential revenue.

Supporting Collaborations for Shared Research and Learning

Utility managers should work with other utilities and collaborations (e.g., WERF) to test and develop new technologies and approaches. Collaborations can pool funds to support university and graduate student research on topics of immediate interest or those with a high potential at the embryonic stage. Utility managers can thereby ensure that data and results from all new technology trials are shared, even when trials are not successful. By working with a variety of other utilities to test and share findings, utility managers will greatly reduce their individual risk and gain greater rewards. One approach may be to establish a "lessons learned" reporting system for objective, independent evaluations that are shared among utilities.

Protecting Utility and Sector with Patents

Patenting innovative processes, once successfully trialed, protects their use for a utility and, if placed in the public domain, provides a resource for the sector. In this way, the public sector can help ensure the free adoption of innovative practices.

Alternative Treatment Technologies

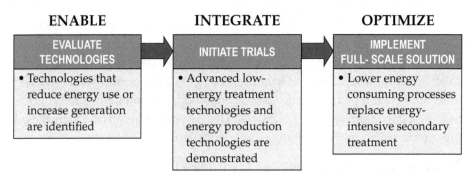

ENABLE	INTEGRATE	OPTIMIZE
EVALUATE TECHNOLOGIES	**INITIATE TRIALS**	**IMPLEMENT FULL- SCALE SOLUTION**
• Technologies that reduce energy use or increase generation are identified	• Advanced low-energy treatment technologies and energy production technologies are demonstrated	• Lower energy consuming processes replace energy-intensive secondary treatment

Energy neutrality in wastewater operations is itself a catalyst for innovation. Recent efforts to consider wastewater treatment facilities as *water resource recovery facilities* and embrace energy neutrality reflect a significant paradigm shift from traditional treatment and disposal. This future vision requires significant technological, operational, and institutional innovation.

Innovative Low-Energy Treatment Technologies

Low-Energy Dewatering

Although drying beds continue to be a staple technology (climatic factors and space permitting), innovative low-energy dewatering approaches are emerging. Solar drying in greenhouse-like enclosures has been used in geographical regions that are not typically associated with passive drying. Use of waste heat from co-generation equipment can be directed to innovative dewatering equipment to aid in drying.

Low-Energy Biological Ammonia Removal

Several processes are in development or in early stages of deployment to provide ammonia removal with substantially lower energy use than conventional nutrient-removal approaches. These systems use specific bacteria that use nitrite (an intermediate product of nitrification) and ammonia to produce nitrogen gas and water (e.g., Anammox).

Microbial Fuel Cells

A microbial fuel cell (MFC) is a device that converts chemical energy to electrical energy by the catalytic reaction of microorganisms. This technology spans both energy-efficient treatment and energy production and may one day be the answer to conventional, energy-intensive aerobic treatment systems. Current research is focused on increasing efficiency and achieving adequate chemical oxygen demand removals. In the future, a utility may be solicited to host an MFC demonstration.

Anaerobic Secondary Treatment

Researchers have been focusing on developing anaerobic treatment systems as a promising solution to minimize energy demand and maximize the design of facilities for nutrient, energy, and water recovery (Kim et al., 2010). Anaerobic systems have significantly less energy demand for processing, create energy-rich biogas, and result in much less biological waste sludge compared to conventional aerobic systems. Currently, there are no full-scale systems operating, but some wastewater leaders believe that solving this challenge is the pinnacle for energy neutrality in the sector.

Innovative Energy Production Technologies
Fats, Oils, and Grease to Energy

Energy production from high-strength organic wastes is increasingly practiced at WRRFs. Organic liquids and residuals, which might otherwise be discharged to sewers with a high energy cost for treatment, may be managed instead for their potential energy value. Of particular interest are FOG, the material pumped from restaurant grease interceptors and traps that poses dilemmas to collection systems but can be intentionally trucked to WRRFs for more efficient energy recovery. Such residuals and FOG may be converted to biogas in municipal anaerobic digesters for use in boilers or generators. Another innovative energy-producing technology recovers low-quality brown grease and converts it to high-quality, American Society for Testing and Materials-compliant biodiesel.

Thermochemical Conversion of Biosolids for Energy

After anaerobic digestion, biosolids still contain a significant amount of energy, which theoretically could be thermochemically converted to a useable form (fuel, heat, and electricity). On a dry basis, the energy content of biosolids is comparable to wood (14 000 to 21 000 kJ/kg). The problem is that biosolids are not dry. Dewatered biosolids still contain 75% or more water, which, during typical thermochemical conversion processes, requires a significant amount of heat (i.e., energy) to convert the solids. Nonetheless, as biosolids reuse or disposal options become more limited and costs rise, utilities are pursuing thermochemical conversion processes to recover the energy directly, create a fuel product (e.g., pellets), or convert biosolids to an alternative renewable transportation fuel.

Technologies range from thermal oxidation (i.e., incineration), in which organics are completely oxidized at high temperatures (800 to 1650 °C) in the presence of excess air, to pyrolysis, in which organics are partially oxidized at high temperatures (400 to 800 °C) in the absence of oxygen. During pyrolysis, the organic material, such as biosolids, is converted to organic compounds, noncondensable gases, and solids. The carbon-based solid product that forms as a result of incomplete combustion is referred as *char* and can be used as a solid fuel alone or mixed with coal as a potential alternative fuel source for power plants. In between thermal oxidation and pyrolysis is gasification, which is the partial oxidation of organics, resulting in the formation of a fuel called *synthesis gas* (carbon monoxide, hydrogen, and methane).

Existing examples of these technologies include a biosolids drying facility in Maryland that produces pellets for use in nearby cement kilns. The pellets replace coal and the residual ash that is created is incorporated to the cement. Another facility in Florida uses heat from a biosolids gasification system to help dry the biosolids. A third facility in California is developing a system to produce high-quality renewable diesel from biosolids.

Algae

Algae recover nutrients from wastewater and yield a biomass with high energy content. Theoretically, the high nutrient content in wastewater is an excellent medium for algae growth, which could allow algae reactors to be used for highly effective, low-energy nutrient removal. Furthermore, algae biomass has high oil content that may be converted to biofuel. Another synergy is that, because algae growth requires carbon dioxide, emissions from a co-generation facility could be exhausted to algae reactors. These factors make algae treatment with energy recovery promising, although no full-scale installations currently exist. Currently, demonstration systems require significant land area and, therefore, are more feasible in rural areas.

Alternative Management Approaches

Decentralized or Hybrid Systems

Not all alternatives to conserve or generate energy are technology related, or they are at least not solved by a new or existing technology. In many instances,

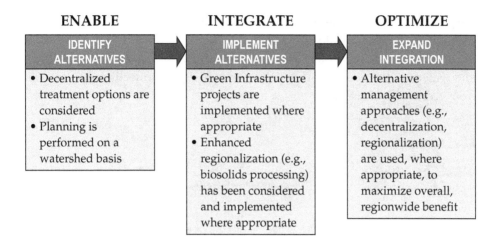

the most efficient means of addressing a problem is by changing the system or management approach. For example, decentralized systems may reduce the need for pumping long distances. Techniques such as wastewater scalping can save energy by removing water from the sewer system, treating it to irrigation quality for local use, and discharging the nutrient load back into the sewer to be handled by the centralized system.

Integrating Innovation and Conservation

One management opportunity is to share the utility's vision for innovation through the entire enterprise. Operators are often unaware of the financial implications of inefficient processes and, instead, are concerned primarily with the goal of meeting effluent limitations. Therefore, many operators (who would otherwise be an excellent source of creative solutions) are not able to contribute because they do not understand the importance or means of reducing energy consumption.

Holistic Vision

Another management option is to engage in a robust examination of the treatment system. Utility managers may find that, with the addition of various effluent limits over time, the traditional response by engineers has been to add additional treatment units. In this incremental manner, utilities may have veered from what might have been an optimal approach for treatment on a holistic scale.

Co-Location and Regionalization

Management leadership may entail looking beyond the fence line for opportunities to develop symbiotic relationships with other industrial activities. Co-location of a power plant and a WRRF can enable the WRRF to use waste heat from a power plant to enhance performance while providing cooling water to the generating plant. Regionalization of solids treatment may be used to obtain a critical mass of organic load for anaerobic digestion to be feasible, especially in rural areas. In some instances, solid waste processors, agricultural feedlots, and water utilities could work together on a co-digestion facility that benefits from economies of scale.

References

Kim, J.; Kim, K.; Ye, H.; Lee, E.; Shin, C.; McCarty, P. L.; Bae, J. (2010) Anaerobic Fluidized Bed Membrane Bioreactor for Wastewater Treatment. *Environ. Sci. Technol.*, **45** (2), 576–581.

U.S. Environmental Protection Agency (2008) *Emerging Technologies for Wastewater Treatment and In-plant Wet Weather Management*; EPA-832/R-06-006; U.S. Environmental Protection Agency: Washington, D.C.

Section 8

Conclusions

Many water utilities are already leading the way to the new model of resource production rather than waste disposal. This transformation provides a real opportunity to benefit communities and the environment while also reducing costs and generating revenue, which benefits ratepayers. As climate change mitigation strategies come into play, the value of renewable energy and energy conservation will continue to increase. Arriving at sustainable energy management requires comprehensive efforts in many areas, as discussed in this document. Nonetheless, efforts in even one or two of these areas can have a significant positive effect on a utility's operations, costs, and culture.

To get started, consider the following tangible targets and how they could be used by a utility:

- Reduce energy use by 20% from baseline within 5 years,
- Eliminate flaring within 10 years, and
- Produce as much renewable energy as the utility consumes within 10 years.

Section 9

Case Studies

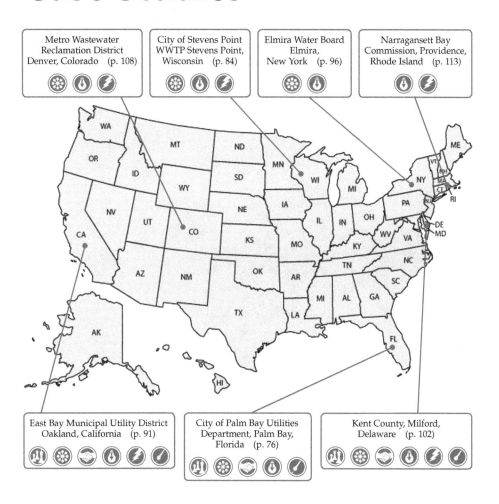

Metro Wastewater Reclamation District Denver, Colorado (p. 108)

City of Stevens Point WWTP Stevens Point, Wisconsin (p. 84)

Elmira Water Board Elmira, New York (p. 96)

Narragansett Bay Commission, Providence, Rhode Island (p. 113)

East Bay Municipal Utility District Oakland, California (p. 91)

City of Palm Bay Utilities Department, Palm Bay, Florida (p. 76)

Kent County, Milford, Delaware (p. 102)

City of Palm Bay Utilities Department, Palm Bay, Florida

Agency	City of Palm Bay Utilities Department
Location	Palm Bay, Florida
Energy Champion Contact	Water: Katie Fought, P.E., Engineering and Plant Operations Division Manager; foughk@pbfl.org Wastewater: Edward Roske, P.E., Engineer II; roskee@pbfl.org

Average Annual Flow

Water: 21.6 ML/d (5.7 mgd)
Wastewater (WW)/Water Reclamation Facility (WRF): 11.5 ML/d (3.0 mgd)

Annual Average Energy Use at Facility (MW)

Water: 0.68
WW/WRF: 0.32

Annual Average Renewable Energy Production at Facility (MW)

Currently none; reviewing alternatives

Annual Average Nonrenewable Energy Production at Facility (MW)

Currently none, to be studied

Type of Treatment
☑ Primary
☑ Secondary
☐ Nutrient Removal

Type of Renewable Energy Source(s)	
☐ Biogas (municipal sludge)	☐ Wind
☐ Biogas (trucked waste)	☐ Geothermal
☐ Solar	☐ Other: _____

Energy Intensity

Water: 760 kWh/ML (2870 kWh/mg)
Wastewater: 670 kWh/ML (2568 kWh/mil. gal)

Utility Overview

The City of Palm Bay Utilities Department (PBUD), located in Palm Bay, Florida, oversees three water treatment plants and one wastewater treatment plant (or water resource recovery facility—WRRF) and one water reclamation facility serving more than 32 000 water and 15 000 sewer customers. The Utilities Department is organized into four divisions (Engineering and Plant Operations; Distribution, Collection, and Maintenance; Business Operations; and Enterprise Geographic Information Systems) under a council-manager form of government.

Energy Goal

Reduce energy costs by 10%. The energy goal was established in Six Sigma Project Charters and uses the baseline calendar year (CY) of 2007 for water facilities and CY 2008 for wastewater/WRRF facilities. Additional energy goals are included in the 2010 Sustainability Master Plan (SMP) (http://www.palmbayflorida.org/citymanager/documents/sustainability_master_plan.pdf).

Progress Toward Goal

A 39.5% reduction in energy costs has been achieved as of CY 2012. PBUD exceeded the SMP goal of a 10% reduction in greenhouse gas (GHG) emissions over a CY 2009 baseline.

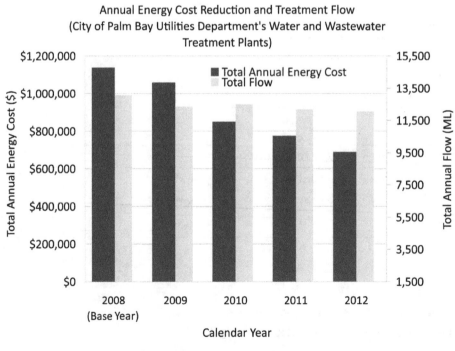

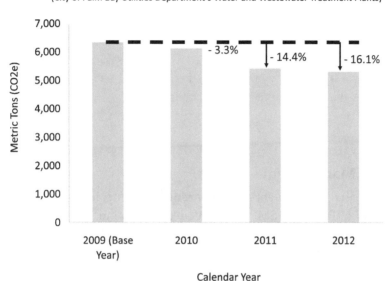

Metric Ton (mt) of Carbon Dioxide Equivalent (CO2e)-Electrical Use
(City of Palm Bay Utilities Department's Water and Wastewater Treatment Plants)

Energy Program Overview

- Environmental Management System (EMS). In 2008, PBUD established its EMS, called "GreenWay", and completed related strategic planning at the city and departmental level. (See http://www.palmbayflorida.org/utilities/about/strategic.html.)
- "Water Treatment Facilities' Energy Consumption, Reduction, and Generation" was identified in the EMS as one of three significant environmental aspects and two Six Sigma projects were implemented to address energy use and reduction at the water and WRRFs. (Energy intensity metric of kWh/1000 gal of treated water was used to normalize and compare across plants.) (See http://www.palmbayflorida.org/utilities/environment/env_aspects.html.)
- In 2010, a City SMP was completed and approved for implementation by the City Council, with the Utilities Department recognized as a leader in sustainability. The 2010 SMP identified activities and projects for the Utilities Department that were relevant and correlated to triple bottom line (TBL) decision making criteria (Environmental, Economic, and Social Equity). (See http://www.palmbayflorida.org/citymanager/documents/sustainability_master_plan.pdf.)
- The 2010 SMP included an Energy Efficiency and Conservation Strategy (Appendix C) and the development of a Governmental Operations Greenhouse Gas Emission Inventory, which used calendar year 2009 as the baseline. The GHG emission reduction goal was, if fiscally possible, to meet Florida's GHG reduction goals as established in Executive Order

07-126. The order called for a 10% reduction in GHG emissions from 2007 levels by 2012, a 25% reduction by 2017, and a 40% reduction by 2025. (See http://gogreen.palmbayflorida.org/PDFs/greenhouse_gas_emission_inventory.pdf.)

- The Utilities Department reduced GHG emissions by 23.7% from the baseline inventory year of 2009 through 2011 as a result of decreased electrical energy and fossil fuel consumption.

Connection to the Energy Roadmap Topic Areas

Strategic Management

- Municipal government and Departmental Strategic Plans supported by a City of Palm Bay SMP with TBL decision-making criteria. These plans have synergistically optimized the planning, budgeting, and successful achievement of energy savings and consideration of alternative energy use.
- Establishment of the municipal government and Utilities Department carbon footprints have been foundational to measuring and meeting GHG emission reduction goals for the Utilities Department's as well as meeting municipal and state GHG reduction goals.

Organizational Culture

- The energy vision for the City of Palm Bay municipal government and Utilities Department was embodied in the 2010 SMP with implementation strategies for green incentives, green standards, education and outreach, and energy management.
- Specific objectives, targets, and programs were initiated and are ongoing in the EMS with Six Sigma projects addressing energy use reduction in treatment processes for both water treatment facilities and WRRFs.
- Green Belt and Black Belt Six Sigma training was provided on a volunteer basis. Six Sigma Project Management training and Six Sigma Champion training was provided to senior management.
- Establishment of and training on a suggestion program and process improvement program (PIP) has empowered employees to make changes.
- The EMS provides employee training, document controls, a corrective and preventive action process, and management review criteria to establish and sustain energy use efficiencies.

Communication and Outreach

- The EMS activities have been branded as "GreenWay" (including a logo, policy, promotional items, Web exposure, press releases, elementary and secondary school outreach programs, etc.).

- Transparency and objectivity are provided through third-party certification of the EMS and sharing of continuous improvement efforts. In response to continuous improvement and risk reduction (business and environmental), Standard and Poor's and Moody's have raised the utilities' water and sewer bond ratings.
- State, regional, and national recognition awards are shared with stakeholders (http://www.palmbayflorida.org/utilities/about/awards.html).
- PBUD was chosen as a case study in the U.S. Environmental Protection Agency Resource Guide to Effective Utility Management and Lean (http://www.epa.gov/lean/environment/pdf/eum-lean-guide.pdf).
- Developed and piloted a water conservation and efficiency educational program ("WaterWise") for 5th graders.
- Two abstracts have been approved for publication as papers and presentations at the Florida Water Resources Conference in April 2013. The titles are "An Innovative and Cost Effective Solution for Improving Your Reclaimed Water Quality" and "The Sky Is Not Falling on Continuing Biosolids Land Application along the Space Coast".

Demand-Side Management

- Six Sigma breakthrough methodologies (define, measure, analyze, improve, and control—DMAIC) required the understanding of electrical billing, electrical use (including peak demand use), treatment flows, processes, and project team management necessary to establish project charters and measurement plans.

- "Biosolids production, handling, and disposal" was targeted in the EMS. A belt filter press installed for biosolids dewatering significantly reduced fuel use and scope 3 GHG emissions by decreasing off-site biosolids hauling by approximately 80%.
- A very successful infiltration and inflow program was initiated within the EMS to reduce treatment energy use and peak demand at the plants and lift stations.
- A supervisory control and data acquisition based power management program that incorporates system capacities, power, and flow monitoring for the entire collection system is being implemented to reduce peak energy use at lift stations and WRRFs.

Energy Generation

- A preliminary analysis has been completed to determine what treatment processes are available based on current fats, oils, and grease and septic component information and the alternative analysis of five revenue and/ or cost-reducing options. Further economic analysis is required.

Innovating for the Future

- The Utilities Department has partnered with a secondary school to support a program that provides classroom, laboratory, and hands-on work experience in water industry career fields. The program is the only one in the state to provide Water and Wastewater Operator License "C" Florida Department of Environmental Protection approved curriculum and pay testing fees for the state licensing exam (http://www.edline.net/ files/_LWILR_/c3624aab1776cc0a3745a49013852ec4/AESWRT_Flyer_ Completed_Draft_12-9-10.pdf).
- Considering production of Class A/AA biosolids to replace current Class B produced through aerobic digestion. Alternatives include chemical treatment, which would eliminate the digester process and $65,400 in net annual electrical costs and $10,100 in net annual maintenance costs.
- The Enterprise GIS (EGIS) Division has reached out to local schools to demonstrate the value of careers in EGIS. The Division developed an interactive presentation based on a survival game that uses zombies to capture the students' interest. The program was highlighted in a major regional newspaper (http://www.floridatoday.com/article/20130106/ NEWS01/301060021/Cram-Session-School-Palm-Bay-use-zombies-help-feed-kids-brains).

Lessons Learned through Implementation of Energy Program

- Implementation of the Energy Program for public utilities must have at its core, transparency, support, and approval of stakeholders (elected officials, advisory boards, senior city management, and rate payers). The City Council approval of municipal and departmental strategic plans, including a comprehensive municipal SMP has been essential for the City of Palm Bay Utilities Department to move ahead with cooperative community support and funding for energy initiatives.
- The training of all employees in environmental awareness, the need for continuous improvement, development of continuous improvement tools, and Lean techniques are essential.
- Training must also include tools for the communication of ideas. A suggestion program should be in place to capture the ideas of all employees for evaluation and a PIP should be in place to document, quantify, and report those suggestions that become improvements to stakeholders.
- *The Energy Roadmap* success achieved to date by PBUD is based on collaborative visioning, planning, and budgeting by stakeholders; a continuous improvement management framework provided by a certified EMS; and a training commitment to all employees by senior management and elected officials who approve departmental line item training budgets.

City of Stevens Point Wastewater Treatment Plant, Stevens Point, Wisconsin

Agency	City of Stevens Point Wastewater Treatment Plant
Location	Stevens Point, Wisconsin
Contact	Jeremy Cramer, Wastewater Superintendent; 715-342-4787; jcramer@stevenspoint.com
Energy Champion Contact	Entire Operations Staff: 715-345-5262; Jeremy Cramer, Wastewater Superintendent; 715-342-4787; jcramer@stevenspoint.com

Average Annual Flow

11.7 ML/d (3.1 mgd)

Annual Average Energy Use at Facility (MW)

0.17

Annual Average Renewable Energy Production at Facility (MW)

0.14

Annual Average Nonrenewable Energy Production at Facility (MW)

0

Type of Treatment
☑ Primary
☑ Secondary
☑ Nutrient Removal

Type of Renewable Energy Source(s)	
☑ Biogas (municipal sludge)	☐ Wind
☑ Biogas (trucked waste)	☐ Geothermal
☐ Solar	☑ Other: Heat pump utilizing wastewater

Energy Intensity

Wastewater Treatment: 345 kWh/ML (1300 kWh/mil. gal)

Utility Overview

The City of Stevens Point, located in Central Wisconsin, owns and operates an 11.7-ML/d (3-mgd) annual average flow water resource recovery facility (WRRF) that serves a population of approximately 27,000 people. The utility also maintains 225 km (140 mi) of sanitary sewer and 15 lift stations. The WRRF was built in 1940 and has undergone two significant upgrades, one in 1972 and another in 1993. The WRRF uses the activated sludge process combined with biological phosphorus removal to meet effluent permit limits and anaerobic digestion to stabilize biosolids. The facility consistently discharges an effluent that averages 5 ppm for biochemical oxygen demand (BOD) and total suspended solids and 0.6 ppm for total phosphorus. Biogas from the anaerobic digesters is burned in a 180-kW engine to produce electricity and hot water. This combined heat and power (CHP) unit currently supplies 84% of the WRRF's electricity demand and almost 100% of the digester heat demand. Excess heat from the CHP unit is also used to maintain temperatures in multiple buildings at the facility during winter months.

To increase biogas production, the facility accepts a wide range of high-strength wastes directly to the anaerobic digesters. High-strength waste addition has doubled daily biogas production and generated significant revenue through tipping fees.

Energy Goal

Energy neutral

Progress Toward Goal

84% energy neutral

Energy Program Overview

For the past 10 years, Stevens Point has focused on being as energy efficient as possible and has emphasized using all equipment and tankage to its maximum potential. When selecting new equipment, energy efficiency and sizing is important. Educating the operations staff on energy usage and electrical demand has been ongoing. Optimization of the different processes used at the facility to consume the least amount of energy has also been important. The energy program used at Stevens Point can be described as a two-step process: energy conservation first, then energy production. The facility uses all electricity produced on site first and then exports additional electricity produced back to the local electrical grid.

Energy Conservation

- Energy awareness
    1. Operations staff educated on electrical usage and demand and how operation of equipment will affect the electrical bill.

2. Daily monitoring of electrical usage and demand.
3. Educating and working with local industries and haulers on how their waste affects the WRRF. (If a waste can be concentrated and put directly into anaerobic digestion, it is then hauled to the facility instead of going down the drain. It increases electricity production and reduces electricity use through aeration.)

- Aeration system upgrades
  1. Installed dissolved oxygen probes to provide more reliable, real-time readings in aeration basins.
  2. Lowered dissolved oxygen level setpoints in the aeration basins from 2.0 to 0.8 mg/L.
  3. Installed new higher-efficiency aeration blower with greater turn-down ability.
  4. Changed from fine-bubble ceramic to fine-bubble membrane diffusers.

- Turned off non-essential equipment
  1. Programmed programmable logic control (PLCs) to prevent running multiple large horsepower motors simultaneously.
  2. Shut off yard lights.
  3. Channel air blowers only run during low flows (1 to 2 hours per day).
  4. Shut off exhaust fans in unused rooms.
  5. Shut off lights in rooms and buildings when not occupied.
  6. Shut down tanks or equipment if not needed.
- Equipment upgrades
  1. Installed lower horsepower plant air compressors that are sized correctly to meet airflow demand for dissolved air flotation, air-operated valves, and air-operated pumps.

2. Upgraded UV light system with automatic cleaning.
3. Reduced horsepower of submersible mixers for the anaerobic/anoxic basins (biological phosphorus removal process).
4. Installed linear motion mixing on anaerobic digestion process.
5. Light-emitting diode lights installed around the plant.

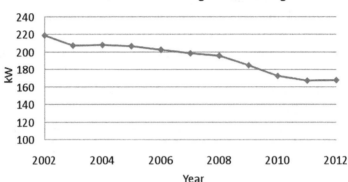

Stevens Point Annual Average Electrical Usage

- Plant heating and cooling upgrades
  1. Installed heat pumps using plant effluent at buildings for heating and cooling.
  2. Installed connection from biogas-fired boiler heat loop to heat multiple buildings.
  3. Heat anaerobic digesters with biogas (boilers or CHP unit).

Energy Production

- Installed a 180-kW biogas engine-generator/CHP unit.
- High-strength waste receiving program
  1. Lowered tipping fees to attract a wide range of waste from haulers.
  2. Worked with haulers on a daily basis to maintain good relationships and attract more feedstock for the digesters.
  3. Installed multiple high- and low-strength waste receiving areas.
  4. Accepted a wide range of wastes, while closely monitoring overall effects to the facility, especially with regard to phosphorus levels.
- Energy produced at the facility is used on site first, then exported to the local electrical grid.

Connection to the Energy Roadmap Topic Areas
Strategic Management

- Benchmark energy consumption and production against similar-sized facilities and set goals.

- Goal of being energy independent.
- Use all equipment and tankage to full potential.
- Understand that we operate an energy production facility as well as a wastewater treatment plant.

Organizational Culture

- Team effort from mayor down to accomplish goal of becoming energy independent.
- Management leads by example and places a great deal of importance on energy conservation.
- Energy use is considered in every decision made regarding equipment purchased, facility upgrades, and process changes.
- Operations staff have embraced focus on energy conservation and production.
- Staff is given the ability to think outside the box and has the ability to be rewarded.
- Very innovative operations staff as a result of the vision of the energy production facility and not just a wastewater treatment facility.

Communication and Outreach

- Worked with local businesses, local waste haulers, and regional waste haulers to attract different feedstock for anaerobic digestion.
- Tours of the facility and presentations given at meetings.

Demand-Side Management

- Electricity use and demand is monitored and analyzed daily, monthly, and annually.
- Historical use and demand have been analyzed and used for the past 10 years to make decisions.
- Electricity billing structure is fully understood and considered in operational decision making.
- Supervisory control and data acquisition monitors usage and production in real time.
- PLC programming reduces demand charges.
- Energy is considered closely in all capital project decision making.
- Collaborated with largest customer (based on BOD loading) to concentrate their wastes for trucking and direct injection to the anaerobic digesters.
- Work with haulers to optimize time of delivery of high-strength waste.

Energy Generation

- Set biogas production goal of 2300 m³/d (80 000 cfd) to run biogas generator at 100% load.

- Secured attractive buy-back rate from local electric utility for electricity exported to the grid.
- Constantly looking for new feedstock for digestion and increased biogas production.

Innovating for the Future

- Staff is aware of existing technologies and is always looking at future possibilities.
- New technologies are piloted at the WRRF.
- In-house trials are conducted for energy saving measures.

Lessons Learned through Implementation of Energy Program

- Size CHP unit and gas treatment correctly (look at all current and future possibilities).
- Feedstock for the digesters is "feast or famine".
- Need to improve high-strength waste receiving and storage to feed digesters properly and optimize biogas production.
- Mixing is needed in high-strength waste receiving station (wastes stratify quickly, especially fats, oil, and grease).
- Slug loading of the digesters changes biogas composition, which greatly affects the biogas cleaning/conditioning system.
- Not all wastes are created equal.
- Waste haulers will introduce a great deal of "unwanted" foreign material to your system.
- Challenges and obstacles will present themselves when trying to move a biogas project forward at a small facility because it is not part of the "normal" treatment process.

East Bay Municipal Utility District, Oakland, California

Agency	East Bay Municipal Utility District (EBMUD)
Location	Oakland, California
Energy Champion Contact	Alicia Chakrabarti, Associate Civil Engineer; 510-287-2059; achakrab@ebmud.com

Average Annual Flow

265 ML/d (70 mgd)

Annual Average Energy Use at Facility (MW)

4.9

Annual Average Renewable Energy Production at Facility (MW)

4.9

Annual Average Nonrenewable Energy Production at Facility (MW)

0

Type of Treatment
☑ Primary
☑ Secondary
☐ Nutrient Removal

Type of Renewable Energy Source(s)	
☑ Biogas (municipal sludge)	☐ Wind
☑ Biogas (trucked waste)	☐ Geothermal
☐ Solar	☐ Other: _____

Energy Intensity

Wastewater Treatment: 440 kWh/ML (1680 kWh/mil. gal)

Utility Overview

East Bay Municipal Utility District is a combined water and wastewater utility serving the greater eastern San Francisco Bay communities of Oakland and Berkeley in addition to smaller surrounding cities. The main facility consists of primary and secondary treatment and disinfection. Wastewater is discharged to the San Francisco Bay.

Energy Goal

Net electricity producer, i.e., renewable electricity production exceeds plant use.

Progress Toward Goal

Exceeding goal by generating renewable electricity at 110% of demand.

Energy Program Overview

Two primary programs:
1. Resource recovery or R2: acceptance of high-strength industrial/commercial wastes for digestion, gas production, and energy generation.

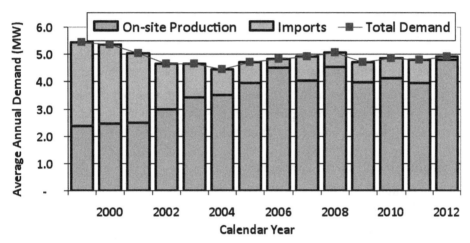

These efforts have increased on-site renewable energy generation by 125% over the last 10 years.

2. Energy system master plan: included energy audit to identify potential areas for conservation. Expect to reduce demand by 10 to 15%. Recently completed in October 2012.

Relationship between Roadmap Topic Areas and Activities at Utility

Strategic Management

- Set goals for energy generation (i.e., become net energy producer) and energy demand reduction (10 to 15% reduction). Identified generation alternatives and used life cycle analysis to select most cost-effective generation.

Organizational Culture

- Form Energy Team and designate leader ("champion"). Energy-related goals incorporated to performance plans. Provide staff training with respect to energy use and demand management.

Communication and Outreach

- Communicate energy goals and successes to customer base. Provide tours to local/regional leaders and news media. Active in information-sharing groups and participate in regulatory advocacy.

Demand-Side Management

- Record and track billing. Conducted audit. Assessed usage against various benchmarks. Enhanced metering capabilities. Manage loading.

Energy Generation

- Set and met generation goals (>100% of demand, "net energy producer"). Identify and manage supplemental digester feedstocks to enhance biogas production. Utilizing renewable energy certificates.

Innovating for the Future

- Actively monitor developments in treatment/generation technologies. Participate in research consortiums (e.g., WERF and local universities). Evaluate alternative technologies and conduct on-site pilot studies to determine suitability.

Lessons Learned through Implementation of Energy Program

It can be difficult to navigate markets for environmental commodities while they are still in development by regulators so prices may fluctuate dramatically and not meet revenue expectations. High-energy feedstocks may disappear as producers find alternative means for disposal or other water resource recovery facilities begin accepting these same wastes. Risk mitigation is important: long-term contracts for supplemental feedstocks can mitigate biogas production risk and long-term energy sales contracts can mitigate pricing/revenue risk.

Elmira Water Board—
Water Treatment Plant,
Elmira, New York

Agency	Elmira Water Board—Water Treatment Plant
Location	Elmira, New York
Energy Champion Contact	Brian Bednarski, P.E., Assistant General Manager; bbednarski@elmirawaterboard.org

Average Annual Flow

25 ML/d (6.5 mgd) (water treatment facility) or
19 ML/d (4.9 mgd) (pumping station)

Annual Average Energy Use at Facility (MW)

0.19 (treatment facility)
0.19 (pumping station) (2007)

Annual Average Renewable Energy Production at Facility (MW)

0

Annual Average Nonrenewable Energy Production at Facility (MW)

0

Type of Treatment	
Drinking Water Treatment	**Wastewater Treatment**
☑ Coagulation/Flocculation	☐ Primary
☑ Clarification	☐ Secondary
☑ Filtration	☐ Nutrient Removal
☑ Pumping	

Type of Renewable Energy Source(s)	
☐ Biogas (municipal sludge)	☐ Wind
☐ Biogas (trucked waste)	☐ Geothermal
☐ Solar	☐ Other: _____

Energy Intensity

Water Treatment: 180 kWh/ML (690 kWh/mil. gal)
Pumping Station: 240 kW/ML (930 kWh/mil. gal)

Utility Overview

The facility is a combined surface water/groundwater treatment facility, where 60% of the influent raw water is from the Chemung River and the remainder is from four different groundwater wells, all fewer than 30 m (100 ft) deep. Historical average flows are in the range of 20 to 30 ML/d (6 to 8 mgd), with higher water demand occurring in the summer because of golf course irrigation. The total service area is 47 km² (18 sq mi), serving an estimated 65,000 people through 18,000 connections in the City of Elmira, New York, and portions of neighboring towns.

The treatment process consists of rapid mix, coagulation, flocculation, clarification, and sand filtration. Treated water is disinfected using chlorine gas and pumped to the finished water reservoir.

Energy Goal

Minimize electric demand and use and reduce associated costs. In many utilities, the electric demand (i.e., instantaneous power) charge can be a significant amount of the bill. The Elmira Water Board has reduced the demand charge to between 10 and 11% of the total electric bill.

Progress Toward Goal

Significant demand charges result when the instantaneous energy use varies considerably. The demand graphs for both the treatment facility and pumping station show that the billed demand is well managed within a narrow range, minimizing demand costs. This results in extremely high load factors of 94.4 and 93.5%, respectively, indicating that both facilities are extremely energy-efficient.

Energy Program Overview

Energy Program:

- Training of operators on demand control—results in operation of non-essential items during off-peak hours (when electricity prices are lower)
- Replacement of motors with premium-efficiency motors
- Installation of variable-frequency drives (VFDs) on equipment that requires variable output
- Use of on-site generator to reduce import demand

Achievements:

- From 2011 Annual Report, "Operating expenses for electric and gas were 85% of the budgeted amount in 2011."
- 2011 Annual Report also noted that the category for pumping and power in 2011 was $6,702 less than in 2010, while selling $98,991 more water.
- Facility has a wire-to-water use of 183 kWh/ML (705 kWh/mil. gal) treated compared with a national average of 370 kWh/ML (1400 kWh/mil. gal) (all facility types), New York State average of 214 kWh/ML (810 kWh/mil. gal) (all facility types), and 214 to 216 kWh/ML (810 to 820 kWh/mil. gal) average for groundwater/surface water facilities.

Variation in Electrical Demand

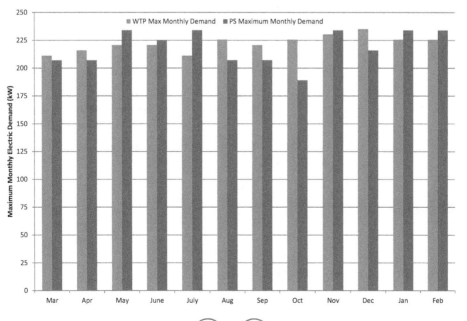

Relationship between Roadmap Topic Areas and Activities at Utility

Strategic Management

- Ongoing leak program to identify and repair leaks reduces pumping costs.
- Well pump motors were found to be oversized; motors were replaced with more appropriately sized motors.

Organizational Culture

- Operators trained and are required to record facility demand on an hourly basis; staff is trained to operate equipment and manage combined demand loads to minimize demand variability. This effort has also resulted in the electric utility granting a "high load factor discount" as the result of having an overall load factor of more than than 67%.
- Standard motors have been replaced with premium efficiency motors.
- Variable-speed drives were installed on many pumps.
- Implementation of successful operator practices and training.
- Training of operations staff on the effect of a variable demand helps keep overall demand in check, minimizing extra demand charges.

Communication and Outreach

None

Demand-Side Management

- Certain activities are restricted to off-peak periods: 29 ML (7.7 mil. gal) of storage is filled. Filters are backwashed. Solids pumping from the clarifiers.
- Facility participates in electric utility load-shifting program to shift electric loads to the 1000-kW diesel generator during high peak demand periods.
- High-pressure cleaning water pump (45 kW, or 60 hp) is operated off of the generator to limit demand and operated manually as required.
- Dewatering operations using belt press performed primarily during off-peak periods.
- The installation of VFDs on finished water pumps resulted in a 20-kW demand per pump reduction.

Energy Generation

None

Innovating for the Future

None

Lessons Learned through Implementation of Energy Program

Education of operators is essential. The Water Board trains staff in how to read the electric bill and how to minimize demand charges. Rewards are offered for keeping demand within a certain range; assistant general manager reports that operators do not want to be the ones that result in high demand. This training allows the operators to make wise decisions on when to operate equipment, especially equipment that is run on an intermittent basis or equipment that is typically not used. For example, water pumps are not used when the backwash pumps are operating. As a result, the facility has consistently maintained overall electrical demand to within a narrow range.

In addition, the utility is proactive in identifying and implementing energy-efficient equipment through the asset management program by installing energy efficient and right-sized motors and equipment.

The facility also participates in the electric utility's demand offset program, which provides discounts for agreeing to shift the facility's loads during high-demand periods on the regional electric grid.

Kent County Regional Wastewater Treatment Facility, Milford, Delaware

Agency	Kent County Regional Wastewater Treatment Facility (KCRWTF)
Location	139 Milford Neck Rd., Milford, DE 19963
Contact	Jim Newton, P.E., BCEE, Environmental Program Manager; 302-335-6000; e-mail: james.newton@co.kent.de.us
Energy Champion Contact	See above

Average Annual Flow

61.7 ML/d (16.3 mgd)

Annual Average Energy Use at Facility (MW)

1.1

Annual Average Renewable Energy Production at Facility (MW)

0.2

Annual Average Nonrenewable Energy Production at Facility (MW)

0

Type of Treatment
❏ Primary
☑ Secondary
☑ Nutrient Removal

Type of Renewable Energy Source(s)	
❏ Biogas (municipal sludge)	❏ Wind
❏ Biogas (trucked waste)	❏ Geothermal
☑ Solar	☑ Other: Heat pump utilizing wastewater

Energy Intensity

Wastewater Treatment: 430 kWh/ML (1620 kWh/mil. gal)

Utility Overview

The KCRWTF is a large, advanced water resource recovery facility that uses the Parkson Biolac system as a biological nutrient removal operation. It serves all

Kent County, Delaware, communities and several communities in New Castle and Sussex Counties. The collection system includes more than 90 pumping and lift stations and more than 800 km (500 mi) of sewer lines. The facility uses an innovative microwave-based UV disinfection process. Biosolids are treated in a filter press with lime stabilization and then either indirectly dried in ovens or solar dried (approximately 20%) in greenhouses that use the Parkson Thermo-System. The biosolids that are oven dried are land applied on local farms, while the solar-dried biosolids are either added to the oven-dried biosolids or are being studied for use as a wood pellet substitute for use in consumer wood stoves. The solar-drying process uses a solar hot water system to heat the concrete pad in greenhouses. The facility's administration building is heated and cooled using a heat exchanger located in the wastewater discharge stream using a quasi-geothermal type process.

Energy Goal

25% renewable electricity by 2015; 30% reduction in natural gas used for biosolids treatment

Progress Toward Goal

15% renewable currently; 8% reduction in natural gas

Energy Program Overview

Energy program was a direct result of the objectives and targets established by the facility's sustainability management system (SMS), which is ISO 14001 certified and uses the Natural Step Framework as its model.

Relationship between Roadmap Topic Areas and Activities at Utility

Strategic Management

- The facility has an SMS that has been certified to ISO 14001, OHSAS 18001, and the National Biosolid Partnership (NBP) Environmental Management System standard since 2006.
- Capital project decision making considers energy savings.
- A greenhouse gas inventory for 2008 through 2011 is used to target energy and chemical use reductions.

Organizational Culture

- All interested parties are kept current on the SMS results.
- A quasi-governmental advisory body (Sewer Advisory Board, or SAB) is the primary mechanism for sewer expansions and the SMS program direction.
- The SMS is governed by a Core Team that sets objectives, targets, and budgets and periodically evaluates progress to ensure that the SMS is continually improved.
- All employees receive annual training on the SMS and progress toward goals.

Communication and Outreach

- The KCRWTF has won a number of regional and national awards for its activities, especially related to its SMS program.
- The governor, state congressmen, and regional U.S. EPA staff have visited the facility and promoted its success.
- The SAB serves as the customer interface for all activities.
- The state regulators and local officials serve as members of the SAB.
- The facility provides tours to regional and national groups and recently hosted more than 300 middle school students studying science and water resources.
- The 2010 Water Environment Federation (WEF) Residuals and Biosolids Specialty Conference included a tour of the new solar dryers.
- Members of the staff are in leadership positions with NBP and WEF.

Demand-Side Management

- The KCRWTF used the U.S. EPA Energy Star Web site to log energy usage and analyze efforts to reduce it.
- The facility participates in a demand response program that allows the facility to reduce the demand on the local grid during peak electrical demand periods by operating its emergency generators.
- A new study is under way to replace the current five 373-kW (500-hp) blowers with a combination of blowers to greatly reduce electric demand.

- The facility has already reduced demand from the blowers by 15% by remotely monitoring dissolved oxygen in the basin and controlling blower operation using this feedback.
- The facility has an active industrial pretreatment program that has won awards from U.S. EPA.

Energy Generation

- The KCRWTF uses the Natural Step Framework as a means to affect sustainable activities.
- The facility has a 1.2-MW photovoltaic solar farm and a solar biosolids drying system and it uses wastewater heat for climate control in the administration building.
- Staff have been monitoring wind resources for 8 years to determine the viability of wind energy, with 5 years of multiple elevation data.
- The facility developed a renewable energy park concept that provided for wind, solar, and hydroelectric power and biomass energy production.
- Hydroelectric and wind resources proved insufficient to be cost effective with current technologies.
- Biomass energy generation would require the construction of anaerobic digesters, which are currently cost prohibitive for the facility.

Innovating for the Future

- KCRWTF has offered to partner with Pennsylvania State University on its microbial fuel cell process by conducting pilot studies.
- The facility is working to develop a new concept for biosolids use by looking at the viability to make the solar-dried biosolids into pellets that could be substituted for wood pellets in consumer wood stoves. The concept is being tested in 2013.
- With the acceptance of land application of biosolids losing favor, alternatives to landfilling are becoming more critical.

Lessons Learned through Implementation of Energy Program

- The heating of the floor in the biosolids dryer is not as efficient as heating the air in the dryer would be. All future solar dryers would use the air heating instead of the floor heating system.
- The photovoltaic solar farm is not large enough to effectively generate all of the necessary electricity needed to run the facility. A future doubling of the system would provide this by banking enough to compensate for times of clouds and night.

Metro Wastewater Reclamation District, Denver, Colorado

Agency	Metro Wastewater Reclamation District
Location	Denver, Colorado
Contact	Wendy Anderson, P.E., Process Engineer; 303-286-3451; e-mail: WAnderson@mwrd.dst.co.us
Energy Champion Contact	Steve Rogowski, P.E., Director of Operations and Maintenance

Average Annual Flow
530 ML/d (140 mgd)

Annual Average Energy Use at Facility (MW)
9.4

Annual Average Renewable Energy Production at Facility (MW)
3.9

Annual Average Nonrenewable Energy Production at Facility (MW)
0

Type of Treatment
☑ Primary
☑ Secondary
☑ Nutrient Removal

Type of Renewable Energy Source(s)	
☑ Biogas (municipal sludge)	❏ Wind
❏ Biogas (trucked waste)	❏ Geothermal
☑ Solar	❏ Other: _____

Energy Intensity
Wastewater Treatment: 426 kWh/ML (1610 kWh/mil. gal)

Utility Overview
Metro Wastewater Reclamation District serves 1.8 million people in the Denver metropolitan area managed by a Board of Directors that represents its 59 municipal connectors. The facility discharges to the effluent-dominated Platte

River, requiring that water resource recovery facilities substantially remove ammonia, nitrogen, and phosphorus.

Energy Goal

The District does not presently have a stated goal.

Progress Toward Goal

15% renewable currently; 8% reduction in natural gas

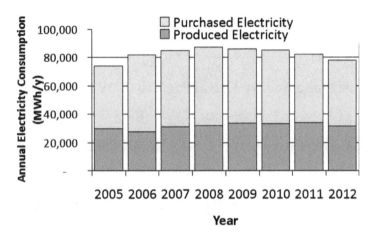

Energy Program Overview

In the last 3 years, an emphasis has been placed on targeting and reducing power consumed for wastewater treatment processes. This has ranged from simply being mindful of removing unnecessary treatment trains from service to implementation of dynamic dissolved oxygen setpoint adjustment based on ammonia-based aeration control.

Relationship between Roadmap Topic Areas and Activities at Utility

Strategic Management

- Provide cost-effective energy management consistent with District goals.

Organizational Culture

- The value of energy management is widely recognized across the District. The District installed its first generation of co-generation facilities more than 20 years ago, when utility rates were still relatively low.
- The District uses a continuous improvement style in its energy program.
- An organizational feature fundamental to the successes of the District's program is its Technical Services Group. These engineering staff are embedded within the operations and maintenance group, are involved in daily activities, and support co-generation and demand-reduction initiatives.

Communication and Outreach

- Participation with Xcel Energy process efficiency initiatives and U.S. Environmental Protection Agency energy conservation initiatives.

Demand-Side Management

- With more than 50% of the connected motor load associated with aeration, Operations staff make process control decisions with aeration energy in mind.
- Standard operating procedures involving large motors limit startup to morning (i.e., low-flow periods) when in-rush is unlikely to set demand peak.
- Increased use of power panels allows real-time monitoring to become a normal routine within daily operation. Performance-based culture challenges Operations staff to take units out of service whenever possible.
- The District conducts rigorous monthly reviews and has a strong understanding of utility billing.

Energy Generation

- Outside service contract is maintained for the operation and maintenance of the District's combined heat and power infrastructure.
- Summer and winter nominations (targets) are developed for co-generation output.
- District has conducted co-digestion trials for evaluation purposes.

Innovating for the Future

- Evaluation of side-stream deammonification for, among other reasons, its reduction in energy requirement.

Lessons Learned through Implementation of Energy Program

Energy markets and costs are dynamic and issues such as tariffs, voter initiatives, sustainability goals, regulations, and technology can influence utility decision making.

Narragansett Bay Commission, Providence, Rhode Island

Agency	Narragansett Bay Commission—Fields Point and Bucklin Point Treatment Facilities
Location	2 Ernest Street, Providence, RI 02905
Energy Champion Contact	Thomas Uva, Director of Policy, Planning & Regulation
	For additional information, please contact Jim McCaughey, Manager of Environmental, Safety & Technical Assistance; 401-461-8848, ext. 352; e-mail: jmccaughey@narrabay.com

Fields Point

Average Annual Flow

151 ML/d (40 mgd) in 2012

Electricity Consumed in 2012 (MW)

1.4

Annual Average Renewable Energy Production at Facility (MW)

0.81 (2013—projected)

Annual Average Nonrenewable Energy Production at Facility (MW)

0

Type of Treatment
☑ Primary
☑ Secondary
☑ Nutrient Removal

Type of Renewable Energy Source(s)	
❑ Biogas (municipal sludge)	☑ Wind
❑ Biogas (trucked waste)	❑ Geothermal
❑ Solar	❑ Other: _____

Bucklin Point

Average Annual Flow

67 ML/d (18 mgd) in 2012

Annual Average Energy Use at Facility (MW)

1.5

Annual Average Renewable Energy Production at Facility (MW)

0.30

Annual Average Nonrenewable Energy Production at Facility (MW)

0

Type of Treatment
☑ Primary
☑ Secondary
☑ Nutrient Removal

Type of Renewable Energy Source(s)	
☑ Biogas (municipal sludge)	☑ Wind
☐ Biogas (trucked waste)	☐ Geothermal
☐ Solar	☐ Other: _____

Energy Intensity

Fields Point: 220 kWh/ML (840 kWh/mil. gal)
Bucklin Point: 540 kWh/ML (2000 kWh/mil. gal)

Utility Overview

The Narragansett Bay Commission is a not-for-profit public corporation formed in 1981 to own and operate the Fields Point Wastewater Treatment Facility. The Narragansett Bay Commission also owns and operates the Bucklin Point Wastewater Treatment Facility that it acquired in 1993. The facilities provide wastewater collection and treatment services to 10 cities and towns. On-site sludge incineration ceased at Fields Point in 2005 when sludge began to be managed as a soil amendment as well as by off-site incineration. Operation of a combined sewer overflow tunnel began in 2008 to reduce wet weather overflows

(Courtesy of Peter Goldberg for Narragansett Bay Commission)

that enter Narragansett Bay. Fields Point will be the largest integrated fixed film activated sludge system, reducing total nitrogen to a concentration of 5 mg/L beginning in 2014.

Energy Goal

By 2015, derive a significant portion of electric consumption (i.e., one-third) from on-site renewable resources and operate the facility at a high efficiency (i.e., top 25th percentile).

Progress Toward Goal

Fields Point: efficiency was in the top 28th percentile in 2012 and the wind turbines begin full production in 2013 and are expected to provide more than one-third of on-site demand.

Bucklin Point: 50% of the way to goal.

Energy Program Overview

An energy-focused environmental management system is used to help control rising energy costs and manage the carbon footprint. Energy use, efficiency, savings, and progress toward goals are tracked at least annually. A general goal is to ensure that energy efficiency is considered when plans are made for the future and that opportunities for improvement are regularly assessed. A short-term goal is to attain operational energy efficiencies that are above average and higher in the long term as determined using U.S. EPA Portfolio Manager. Potential for using renewable energy resources has been assessed at each facility and prioritized based on feasibility. Wind turbines were installed in 2012 at Fields Point and other renewable generation is being studied and designed at the Bucklin Point facility.

Connection to the Energy Roadmap Topic Areas

Strategic Management

Strategic Direction

- Benchmarking the normalized energy consumption on an overall basis and at the process level, where feasible.
- Goal setting with regard to U.S. EPA Portfolio Manager rating and on-site production of renewable energy.
- Agency-wide strategic plan to direct performance-based management.
- Energy program uses a plan/do/check/act process similar to ISO 14000 Environmental Management System.
- Uses triple bottom line reasoning in decision making.
- Active role educating state legislators and employs a full-time Legislative Liaison.

Financial Viability

- Economic feasibility is based on life cycle analysis that assesses return on investment and not just payback period.
- Request for bids on projects, renewable energy credits, and demand response contracts are based on maximizing the return on investment or revenue for the agency.

Collaborative Partnerships

- Established partnerships throughout Rhode Island (Renewable Energy Fund, Office of Energy Resources, and the General Assembly) for education and advocacy.

Toward Carbon Neutrality

- Annual accounting of all significant contributions to the carbon footprint of both facilities, including fugitive emissions.
- Carbon credits considered in assessing options for managing biosolids.

Organizational Culture

Energy Vision

- Narragansett Bay Commission Division Director and Energy Champion, Thomas Uva, developed Commission's vision of operating highly efficient water resource recovery facilities eligible to receive a U.S. EPA Energy Star award (as available) and using renewable energy in the form of Rhode Island's first wind farm and ultimately becoming a net energy exporter.

Energy Team
- Key employees work cooperatively on a multidisciplinary energy team.
- Energy team tracks energy consumption, efficiency, savings, and progress toward preset goals and reports findings at least annually to management and staff.

Communication and Outreach
Customers and Community
- Proactively surveyed 300 neighbors of the proposed wind turbine project (Fields Point).
- Work with restaurants to promote best energy management practices for fats, oils, and grease.

Regulatory and Legislative
- Early involvement of key stakeholders is invaluable with respect to complying with existing building regulations, identifying air emissions restrictions, and gaining public support.
- Staff advocates at the state level for renewable energy financial incentives and reasonable siting guidelines.

Media Outreach
- Developed high-quality media kit (video, photographs, and a Web site).
- Public relations staff works on messaging with media.
- Regular tours.

Demand-Side Management
Electricity Costs and Billing
- Utility bills are reviewed and tracked.
- Energy team members educate others on understanding energy bills.
- Participate in demand response and curtailment using emergency generators (since 2007).

Power Measurement and Control
- Overall and process-level energy consumption is tracked and compared to baselines and benchmarks that have been determined.
- Real-time submetering and control are in place for total electric use and use by significant process operations.

Energy Management
- Energy assessments and audits are conducted at least once per year.
- Energy savings are designed into capital projects.
- Consider turning off non-essential equipment for short periods of time when maximum demand approaches a given threshold.

Source Control
- Established a goal to distribute daily centrate loading to biological nutrient removal (BNR) processes.
- Ongoing infiltration and inflow program.

Energy Generation
Strategy
- Short-term goal to have renewable generation compose one-third of on-site consumption by 2015.

Energy from Water and Wastewater
- Prioritized renewable sources include solar, wind, biogas, and hydroelectric power.
- Biogas is generated and used to provide heat to offset natural gas costs (Bucklin Point).
- Possible uses for excess biogas-derived heat generated in the summertime are being investigated.
- Optimize biogas production by feeding digesters at a more frequent and constant rate.
- Digester mixing is maintained at an adequate level and digesters are on a schedule for degritting.
- Primary sludge thickening was improved using variable-frequency drive control for sludge pumps.
- Waste activated sludge thickening is being improved by replacing an old dissolved air flotation system with a new belt press.

Supplemental Energy Sources
- Assessed co-digestion of supplemental waste (Bucklin Point).
- Considered accepting waste deicing glycol fluid from nearby airports.
- A study of the existing anaerobic digestion systems design and current operating parameters revealed that the influent volatile solids loading rate was already at its design value and its proximity to the maximum allowable loading rate left few options for co-digestion.
- A demonstration project was considered using a single digester that would accept engineered food waste slurry. However, no means to economically

use additional heat resulting from the supplemental waste. Considered cleaning excess biogas for introduction to the gas utility supply line.

- Preengineered food waste was found to contain nitrogen that, if released in the digester, was estimated to increase the concentration of total nitrogen in the final treated effluent.
- Co-digestion options may possibly be revisited when newly upgraded BNR process demonstrates its can meet the new total nitrogen limit.

Renewable Energy Certificates

- Wind turbines are certified producers of class 1 renewable energy credits in Rhode Island and Massachusetts.
- Energy Team members from the Commission's Finance group are responsible for securing the maximum value for renewable energy credits.

Innovating for the Future

Research and Development

- Staff keep abreast of emerging or innovative methods to reduce the carbon footprint.
- Seek grants or other funding sources to reduce risk for emerging opportunities.
- State-of-the-art analytical laboratory provides key analyses to further in-house research.

Risk Management

- Worked collaboratively to develop an internal risk management procedure/policy with regard to the danger of ice thrown from wind turbines.
- Bucklin Point will continue to work on assessing ways to minimize risk associated with overloading the digesters with volatile solids. After 2013, the facility may continue exploring ways to boost biogas production with minimal nitrogen effects.

Alternative Technologies

- Advancements in emerging technologies capable of reducing the carbon footprint are closely followed, including microbial fuel cell research, carbon dioxide sequestration, side-stream treatment, bioaugmentation to reducing supplemental carbon, anaerobic ammonia oxidation, primary sludge fermentation, and enhanced waste activated sludge digestion.
- The Commission's history illustrates a willingness to host demonstration projects that advance beneficial technologies and reduce risk, for example, by piloting the use of wetlands for wastewater treatment and installing

green low-impact development technologies at its new Administration Building.

- Discussed regional anaerobic digestion facility with interested developers.

Alternative Management Approaches

- The Fields Point facility will meet a total nitrogen limit of 5 mg/L starting in 2014 with the largest fixed-film activated sludge process.
- Highly efficient turbo blowers were installed to meet the increased demand for air and primary solids removal was increased.
- The Commission continues to use its own staff to analyze receiving water quality and make improvements on a watershed-wide basis.
- Considered decentralized stormwater treatment approaches.
- The Commission is urging Rhode Island regulators to consider fair watershed management approaches to nitrogen removal using a total maximum daily limit approach.

Lessons Learned through Implementation of Energy Program

A formal strategic plan and mission statement are necessary to manage energy and maintain a leadership role. In addition, an effective energy team helped overcome challenges and facilitated progress toward attainment of goals.

An energy-focused environmental management system considers economics as well as the environment. This system supports the Commission's mission of providing safe and reliable services at a reasonable cost and helps stabilize future rate increases while stimulating the local economy.

Test Drives

Madison Metropolitan Sewerage District, Madison, Wisconsin

Agency	Madison Metropolitan Sewerage District
Location	Madison, Wisconsin
Energy Champion Contact	Alan Grooms, Process & Research Engineer; 608-222-1201, ext. 253; e-mail: alang@madsewer.org

Average Annual Flow

151 ML/d (40 mgd)

Annual Average Energy Use at Facility (MW)

2.9

Annual Average Renewable Energy Production at Facility (MW)

0.6

Annual Average Nonrenewable Energy Production at Facility (MW)

0.2

Type of Treatment
☐ Primary
☑ Secondary
☑ Nutrient Removal

Type of Renewable Energy Source(s)	
☑ Biogas (municipal sludge)	☐ Wind
☐ Biogas (trucked waste)	☐ Geothermal
☐ Solar	☐ Other: _____

Energy Intensity

Wastewater Treatment (including pumping): 460 kWh/ML (1740 kWh/mil. gal)

Utility Overview

Status of Energy Program

At the Nine Springs Wastewater Treatment Plant, Madison Metropolitan Sewerage District has been capturing and using biogas from anaerobic digestion since the 1930s. Currently, Nine Springs uses all produced biogas to fuel two induction generators for electrical power generation and heat, a single biogas engine to power an aeration blower, and a hot water boiler. Water heated in the boilers and waste heat recovered from engine cooling and exhaust heat recovery are used to heat digesters as well as provide building heat and some air conditioning (via an adsorption chiller system).

Plans/Intention for Energy Program

Ultimately, the objective for the District is to attain energy independence for the main facilities (Nine Springs, the collection and pumping system, and the vehicle fleet). A project (Energy Baseline and Optimization Roadmap) is kicking off in January 2013 with the aid of a consultant to document the current baseline energy condition and create a high-level planning document to frame and guide near- and long-term decisions as a course toward energy independence is charted.

Energy Goal

Attain energy independence

Progress Toward Goal

Approximately 30 to 35% (including heat recovery)

1. **Current level of importance and achievement in six topic areas. Assign a value of 1 (lowest) to 5 (highest).**

Topic area	Importance	Achievement
Strategic Management	4	3
Organizational Culture	3	2
Communication and Outreach	3	1
Demand-Side Management	5	3
Energy Generation	4	3
Innovating for the Future	4	3

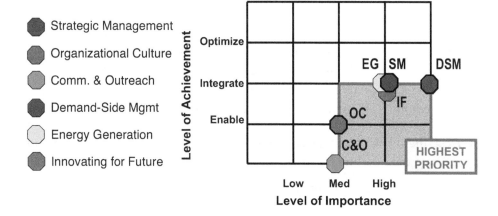

2. **Indicate the areas you plan to focus on as you develop your energy program.**

   - Demand-Side Management,
   - Strategic Management, and
   - Organizational Culture (in that order).

   It is anticipated that all areas will receive some attention.

3. **What ideas from *The Energy Roadmap* were more useful in terms of actually implementing at your utility?**

   Probably the considerations outlined in the areas of Communication and Outreach and Organizational Culture. While we are aware of the importance of these areas on some level, we agree that much more can be done here and the ideas extended in these categories will feed additional tailoring, thought, and innovation as it pertains to our specific situation.

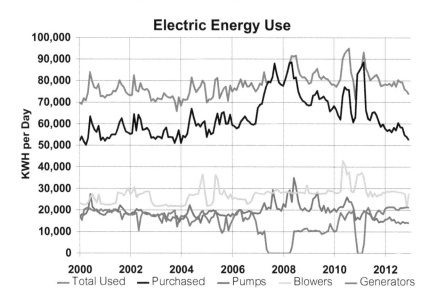

4. What ideas from *The Energy Roadmap* were most surprising?

The degree of emphasis on "soft" changes, such as Organizational Culture and Communication and Outreach. While these areas make sense, they are not as often discussed as more tangible areas such as Demand-Side Management or Energy Generation. Yet, without changes in attitude, culture, and communication, the best ideas will fail for lack of support.

5. Describe the actions that you took to develop your energy program and indicate the timeline for each.

Our energy program is really just beginning to kick off, but actions taken to date include (in no particular order): (1) accepting "brainstorm" suggestions from anyone inside the organization; (2) increasing awareness of the ultimate objective we are setting as well as our current state with respect to it; and (3) open discussion of possibilities and challenges so that the interrelationship of each decision begins to be appreciated. These actions provide some ideas for demand reduction but also serve to change our organizational culture in that we begin to appreciate the bigger picture of where we are trying to go.

6. Describe the results (or expected results) of the actions taken and timeline for each.

With the initial energy study and baseline documentation effort just kicking off, we anticipate coming out of that with a firm idea of where we are right now so that we can measure progress. We also expect that we will continue to change organizationally with our attitudes and an increased focus on how our seemingly minor decisions and actions can affect the whole, either positively or negatively.

South Truckee Meadows Water Reclamation Facility, Reno, Nevada

South Truckee Meadows Wastewater Facility

Agency	Washoe County Community Services Department
Location	South Truckee Meadows Water Reclamation Facility
Energy Champion Contact	John Hulett, Senior Environmental Engineer; 775-954-4612; e-mail: jhulett@washoecounty.us

Average Annual Flow

13.1 ML/d (3.5 mgd)

Annual Average Energy Use at Facility (MW)

Total =	**0.53**
Lift Station =	0.02
Plant =	0.32
Reuse =	0.19

Annual Average Renewable Energy Production at Facility (MW)

0

Annual Average Nonrenewable Energy Production at Facility (MW)

0

Type of Treatment
☑ Primary
☑ Secondary
☑ Nutrient Removal

Type of Renewable Energy Source(s)	
☑ Biogas (municipal sludge)	☐ Wind
☐ Biogas (trucked waste)	☐ Geothermal
☐ Solar	☐ Other: _____

Energy Intensity

Wastewater Treatment (without reuse): 620 kWh/ML (2300 kWh/mil. gal)
Reuse: 350 kWh/ML (1320 kWh/mil. gal)
Total Combined: 970 kWh/ML (3620 kWh/mil. gal)

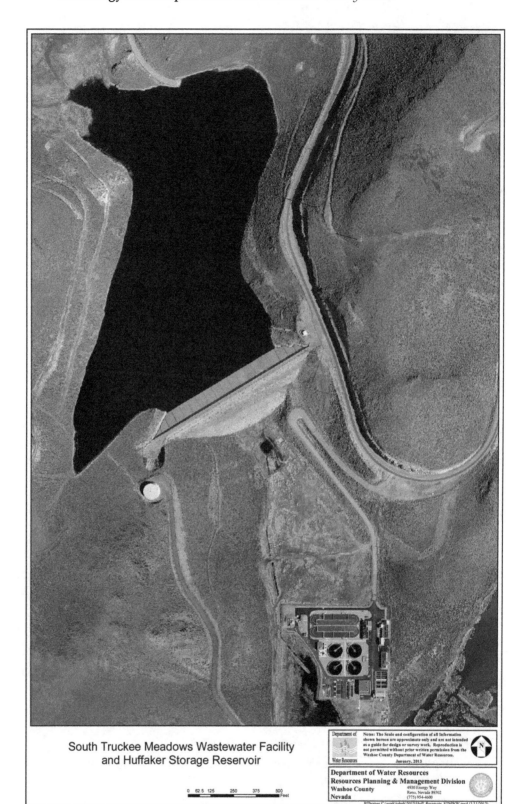

South Truckee Meadows Wastewater Facility
and Huffaker Storage Reservoir

Utility Overview

Washoe County's Department of Water Resources operates the South Truckee Meadows Water Reclamation Facility. South Truckee Meadows is permitted to treat up to 17 ML/d (4.5 mgd), while filtration and chlorine disinfection unit processes are sized for 23 ML/d (6 mgd). Influent flows are presently 12 ML/d (3.2 mgd). The facility achieves partial nitrification/denitrification (typically less than 7 mg/L total nitrogen), with an extended aeration process. South Truckee Meadows' effluent is used entirely for surface irrigation, with no discharge to a receiving body. During nonirrigation periods (November through March), effluent is stored in a 4.9 mil. m³ (4000 ac-ft) reservoir.

Status of Energy Program

1. Integration of control system and power management to shift/shape loads to reduce costs based on time-of-use rate structure.
2. Power surveys to compile records of energy usage at the facility to prioritize those areas yielding greatest savings.
3. Investigating several process alternatives for on-site energy generation, solids reduction, solids drying, and dewatering.
4. Established in-plant water quality and process control objectives.
5. Initiated electrical equipment review and testing plan, which identifies equipment for replacement with higher efficiency units or possible variable-frequency operation.
6. Initiated a unit process modeling program with the University of Nevada Civil and Environmental Engineering Department.

Plans/Intention for Energy Program

Describe where you going with your energy program in short and long term, as known.

The next step in the energy program is to establish effluent and reuse water quality goals and use modeling to determine how best to balance water quality and energy objectives. The County has experienced intermittent water quality problems within the large open reservoir that is used to store water before distribution to 250 reuse customers. The County has recently initiated a modeling project (using BioWin by EnviroSim Associates Ltd., Hamilton, Ontario, Canada) to better understand how aeration (energy usage) affects water quality (nitrate and ammonia in effluent).

Energy Goal

Goals based on cost per volume treated and energy use per volume treated will be established in July 2013.

Progress Toward Goal

N/A

The Washoe County Community Services Department has recently undertaken an evaluation of energy sustainability at South Truckee Meadows using *The Energy Roadmap*. While many elements of the roadmap are priorities for the utility, such as those related to demand-side management and innovating for the future, some elements are not likely to be implemented, such as those related to communication and outreach.

1. **Current level of importance and achievement in six topic areas. Assign a value of 1 (lowest) to 5 (highest).**

Topic area	Importance	Achievement
Strategic Management	3	2
Organizational Culture	4	2
Communication and Outreach	2	1
Demand-Side Management	5	4
Energy Generation	3	3
Innovating for the Future	5	4

Prioritization Graphic – Truckee Meadows

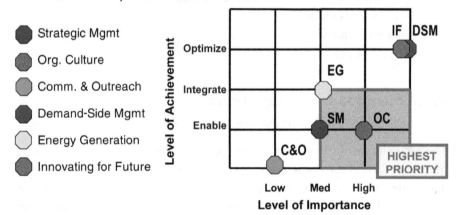

2. **Indicate the areas you plan to focus on as you develop your energy program.**

   The initial focus will be on Strategic Management
   a. Identify energy usage required to meet new internal treated effluent water quality objectives.
   b. Modernize the facilitiy's electrical equipment and process control software.
   c. Implement more regular testing, monitoring, and replacing of high-energy usage equipment, such as the fine-bubble diffuser panels.

3. **What ideas from *The Energy Roadmap* were more useful in terms of actually implementing at your utility?**

   a. Created an "Energy Team" to bring together our staff of operators, engineers, and managers.
   b. Broadened our viewpoints and elevated the work purpose; provided the team a true "energy" project, not simply something to look at when we had time to investigate efficiencies.
   c. Helped us to reprioritize capital improvement plan projects—schedules have been accelerated for projects that focus on energy management or energy conservation (revenue).
   d. Helped us set water quality and energy usage goals.
   e. Added value (new dimension) when evaluating process changes or equipment selection.

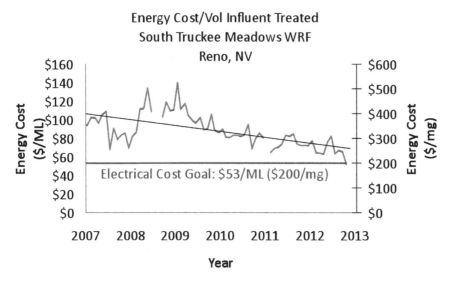

4. **What ideas from *The Energy Roadmap* were most surprising?**

   a. Having the draft Matrix helped focus our efforts to see that incremental steps are important.
   b. It seems very clear that *The Energy Roadmap* can be extremely useful to help utilities with initial the steps of an energy program and refocusing/ refining established programs.
   c. Realization that many utilities likely are far ahead of Washoe County in strategic thinking and organizational culture, areas where we scored ourselves lowest.
   d. Likewise, establishing carbon credits and renewable energy credits are potential future opportunities (note that the region does have a small amount of renewable energy credits from South Truckee Meadows' co-generation facilities).

5. **Describe the actions that you took to develop your energy program and indicate the timeline for each.**

- Power survey—12 months to observe a full year of plant operation.
- Control system upgrade—12 months to fully implement, including adding an "Energy Dashboard" (within 6 months).
- Set water quality objectives (beyond permitted values)—first month. Installed new dissolved oxygen and nitrate probes in aeration basins (completed).
- Developed process modeling to help manage future energy/water quality improvement projects (complete within next 3 months).
- Developed a plan for electrical usage/electrical equipment life cycle and efficiency to develop an asset management/equipment replacement schedule.
- Completed a solids management study to investigate options for energy/nutrient management.

6. **Describe the results (or expected results) of the actions taken and timeline for each.**

Shifting to a time-of-use rate structure provided significant savings. Controls project has also had a significant effect on recent power costs (past 6 months). Overall plant efficiency has increased through pumping strategy optimization (return activated sludge and effluent pumping). Some of the cost savings are also a result of a slight decline in rates charged by the local electric utility.

The Energy Roadmap Matrix provides a detailed framework, which outlines various strategies to become energy sustainable or refocus/refine established energy programs. The County discovered that following the framework of *The Energy Roadmap* provides a means to purposefully bring operators, engineers, and managers together to undertake a common goal, which will bring tremendous value to the facility and utility customers.

Section 10

Additional Resources

Strategic Management

Database of State and Utility Incentives for Renewables and Efficiency. http://www.dsireusa.org (accessed Dec 2012).

National Association of Energy Service Companies. http://www.naesco.org (accessed Dec 2012).

U.S. Environmental Protection Agency (2008) *Ensuring a Sustainable Future: An Energy Management Guidebook for Wastewater and Water Utilities*. U.S. Environmental Protection Agency: Washington, D.C.

Energy Generation

Database of State and Utility Incentives for Renewables and Efficiency. http://www.dsireusa.org (accessed Dec 2012).

Science Applications International Corporation (2006) *Water and Wastewater Industry Energy Best Practice Guidebook*; Prepared for the State of Wisconsin; Focus on Energy: Madison, Wisconsin.

Communications

U.S. Environmental Protection Agency, Landfill Methane Outreach Program, Marketing & Communications Toolkit. http://www.epa.gov/lmop/partners/toolkit/ (accessed Jan 2013).

Utility Branding Network. *Building the Wastewater Utility Brand: Practical Advice for Increasing Trust, Support, and Investment.* http://utilitybranding.net/ (accessed Jan 2013).

Water Environment Federation (2012) *Sustainability Reporting Statements for Wastewater Systems*; Special Publication; Water Environment Federation: Alexandria, Virginia.

Implementing Energy Management Programs

U.S. Environmental Protection Agency, *Energy Management Guidebook and Planning for Sustainability Handbook*. http://water.epa.gov/infrastructure/sustain/cut_energy.cfm (accessed Jan 2013).

Water Environment Research Foundation (2010) *Overview of State Energy Programs for the Wastewater Sector*; Project No. OWSO6R07b; Water Environment Research Foundation: Alexandria, Virginia.

General Energy and Management

American Council for an Energy-Efficient Economy, Addressing the Energy-Water Nexus: A Blueprint for Action and Policy Agenda. http://www.aceee.org/white-paper/addressing-the-energy-water-nexus (accessed Jan 2013).

Water Effective Utility Management. Attributes of Effectively Managed Utilities. http://www.watereum.org (accessed Jan 2013).

Water Environment Research Foundation (2010) *Technology Roadmap to Sustainable Wastewater Treatment in a Carbon-Constrained World*; Project No. OWSO4R07d; Water Environment Research Foundation: Alexandria, Virginia.

Water Environment Research Foundation (2011) *Decision Support System for Sustainable Energy Management*; Project No. OWSO7C07; Water Environment Research Foundation: Alexandria, Virginia.

Water Environment Research Foundation (2011) *Energy Production and Efficiency Research—The Roadmap to Net-Zero Energy*; Water Environment Research Foundation: Alexandria, Virginia.

Partner Identification

National Association of Energy Service Companies. http://www.naesco.org (accessed Jan 2013).

U.S. Department of Energy, Database of State Incentives for Renewables & Efficiencies. http://www.dsireusa.org (accessed Jan 2013).

Benchmarking, Energy Use, and Optimization

AECOM (2011) *National Water & Wastewater Benchmarking Initiative (NWWBI) Public Report*; AECOM: Vancouver, Canada.

Crawford, G. V. (2010) *Best Practices for Sustainable Wastewater Treatment*; WERF Project No. OWSO4R07a; Water Environment Research Foundation: Alexandria, Virginia.

Duke Energy, Understanding Demand and Consumption. http://www.think-energy.net/KWvsKWH.htm (accessed Jan 2013).

New York State Energy Research and Development Authority (2008) *Statewide Assessment of Energy Use by the Municipal Water and Wastewater Sector*; Final Report 08-17; New York State Energy Research and Development Authority: Albany, New York.

New York State Energy Research and Development Authority (2010) *Water and Wastewater Energy Management Best Practices Handbook*; New York State Energy and Research Development Authority: Albany, New York.

New York State Energy Research and Development Authority, Tools and Materials. http://www.nyserda.ny.gov/Commercial-and-Industrial/Sectors/Municipal-Water-and-Wastewater-Facilities/MWWT-Tools-and-Materials.aspx?sc_database=web (accessed Jan 2013).

O'Connor, K.; Yonkin, M. (2009) *Statewide Assessment of Energy Use by the Municipal Water and Wastewater Sector*; New York State Energy Research and Development Authority: Albany, New York.

Pakenas, L. (1995) Energy Efficiency in Municipal Wastewater Treatment Plants; New York State Energy Research and Development Authority: Albany, New York.

Science Applications International Corporation (2006) *Water and Wastewater Industry Energy Best Practice Guidebook*; Prepared for the State of Wisconsin; Focus on Energy: Madison, Wisconsin.

U.S. Environmental Protection Agency (2008) *Ensuring a Sustainable Future: An Energy Management Guidebook for Wastewater and Water Utilities*; U.S. Environmental Protection Agency: Washington, D.C.

Water Environment Research Foundation (1998) *Biosolids Management: Evaluation of Innovative Processes*; Project No. 96-REM-1; Water Environment Research Foundation: Alexandria, Virginia.

Water Environment Research Foundation (2009) *Integrated Methods for Wastewater Treatment Plant Upgrading and Optimization*; Project No. 04-CTS-5; Water Environment Research Foundation: Alexandria, Virginia.

Water Environment Research Foundation (2010) *Energy Efficiency in Value Engineering: Barriers and Pathways*; Project No. OWSO6R07a; Water Environment Research Foundation: Alexandria, Virginia.

Water Environment Research Foundation (2011) *Optimizing Biotreatment Integrating Process Models and Control Technology*; Project No. 03-CTS-11; Water Environment Research Foundation: Alexandria, Virginia.

Innovative Approaches

American Energy Innovation Council (2011) *Catalyzing American Ingenuity: The Role of Government in Energy Innovation*; American Energy Innovation Council: Washington, D.C.

Guest, J.; Skerlos, S.; Barnard, J.; Beck, M.; Daigger, G.; Hilger, H.; Jackson, S.; Karvazy, K.; Kelly, L.; Macpherson, L.; Mihelcic, J.; Pramanik, A.; Raskin, L.; Van Loosdrecht, M.; Yeh, D.; Love, N. (2009) A New Planning and Design Paradigm to Achieve Sustainable Resource Recovery from Wastewater. *Environ. Sci. Technol.*, **43** (16), 6126–6130.

Rubino, V.; Katehis, D.; Sharp, R.; Dailey, S.; Daigger, G.; Young, P.; Deur, A.; Beckmann, K. (2010) Managing Innovation: Optimizing Resource Allocation Using New York City's Innovative Technology Prioritization Tool. *Proceedings of the 83rd Annual Water Environment Federation Technical Exhibition and Conference* [CD-ROM]; New Orleans, Louisiana, Oct 2–6; Water Environment Federation: Alexandria, Virginia.

Water Environment Research Foundation (2011) *Demonstrating Power Production in Wastewater Treatment Processes on a Pilot-Scale Basis*; Project No. OWSO8R09; Water Environment Research Foundation: Alexandria, Virginia.

Tools

Water Environment Research Foundation, Carbon Heat Energy Assessment and Plant Evaluation Tool (CHEApet). http://cheapet.werf.org (accessed Jan 2013).

Water Environment Research Foundation, Life Cycle Assessment Manager for Energy Recovery (LCAMER) Tool. http://www.werf.org/lcamer (accessed Jan 2013).

Water Environment Research Foundation, Green Energy Life Cycle Assessment Tool (GELCAT) and User Manual. http://www.werf.org/a/ka/Search/ResearchProfile.aspx?ReportId=OWSO6R07c (accessed Jan 2013).

Technology Resources

National Renewable Energy Laboratory Web site. http://www.nrel.gov/analysis/pubs_solar.html (accessed Jan 2013).

Water Environment Federation (2010) *Energy Conservation in Water and Wastewater Facilities*; Manual of Practice No. 32; McGraw-Hill: New York.

Water Environment Research Foundation (2010) *Energy Efficiency in Wastewater Treatment in North America: A Compendium of Best Practices for Energy Efficiency and Recovery*; Project No. OWSO4R07e; Water Environment Research Foundation: Alexandria, Virginia.

Water Environment Research Foundation (2010) *Co-Digestion of Organic Waste Products with Wastewater Solids—Interim Report*; Project No. WSO5R07a; Water Environment Research Foundation: Alexandria, Virginia.

Water Environment Research Foundation (2011) *State of the Science on Biogas*; Project No. OWSO10C10a; Water Environment Research Foundation: Alexandria, Virginia.

Technical Information

Daigger, G. (2009) Evolving Urban Water and Residuals Management Paradigms: Water Reclamation and Reuse, Decentralization, and Resource Recovery. *Proceedings of the 82nd Annual Water Environment Federation Technical Exhibition and Conference* [CD-ROM]; Chicago, Illinois, Oct 18–22; Water Environment Federation: Alexandria, Virginia.

Parry, D.L. (1991) A Second Law Analysis of Biogas Utilization Systems for Wastewater Treatment Plants. *Proceedings of the* American Society of Mechanical Engineers/Japan Society of Mechanical Engineers *Thermal Engineering Joint Conference*; American Society of Mechanical Engineers: New York; Vol. 3.

Water Environment Research Foundation (2008) State of the Science Report Energy and Resource Recovery from Sludge; Project No. OWSO3R07; Water Environment Research Foundation: Alexandria, Virginia.

Whitlock, D.; Daigger, G.; McCoy, N. (2007) The Future of Sustainable Water Management: Using a Value Chain Analysis to Achieve a Zero Waste Society. *Proceedings of the 80th Annual Water Environment Federation Annual Technical Exhibition and Conference* [CD-ROM]; San Diego, California, Oct 13–17; Water Environment Federation: Alexandria, Virginia.

Case Studies

Water Environment Research Foundation (2010) *Best Practices for Sustainable Wastewater Treatment*; Project No. OWSO4R07a; Water Environment Research Foundation: Alexandria, Virginia.

Water Environment Research Foundation (2010) *Sustainable Treatment Case Study—Evaluation of the Strass in Zillertal WWTP*; Project No. OWSO4R07b; Water Environment Research Foundation: Alexandria, Virginia.

Water Environment Research Foundation (2011) *Demonstration of the Carbon, Heat, Energy Assessment Plant Evaluation Tool (CHEApet)*; Project No. OWSO4R07g; Water Environment Research Foundation: Alexandria, Virginia.

Water Environment Research Foundation (2011) Energy Efficiency in the Water Industry: A Compendium of Best Practices and Case Studies—A Global Report; Project No. OWSO9C09; Water Environment Research Foundation: Alexandria, Virginia.

Water Environment Research Foundation (2011) *Site Demonstration of the Life Cycle Assessment Manager for Energy Recovery (LCAMER)*; Project No. OWSO4R07f; Water Environment Research Foundation: Alexandria, Virginia.

Appendix

The Energy Roadmap: A Water and Wastewater Utility Guide to More Sustainable Energy Management **and** ISO 50001—Energy Management

Overview of ISO 50001—Energy Management

This standard, published by the International Organization for Standardization in 2011, is applicable to all types and sizes of organizations, regardless of geographical, cultural, or social conditions. It is appropriate for use by the water sector and complements *The Energy Roadmap*. ISO 50001—Energy Management is designed to be used either independently or in alignment with another management system, such as this guidance document. Successful implementation of *The Energy Roadmap*, ISO 50001, or both depends on commitment from all levels and functions of the organization, especially top management.

The purpose of ISO 50001 is to enable organizations to establish the systems and processes necessary to improve energy performance, including energy efficiency, use, and consumption. The standard is intended to lead to reduced environmental effects (including reductions in greenhouse gas emissions) and energy cost savings through systematic management of energy.

ISO 50001 specifies requirements for an energy management system upon which an organization can develop and implement an energy policy and

141

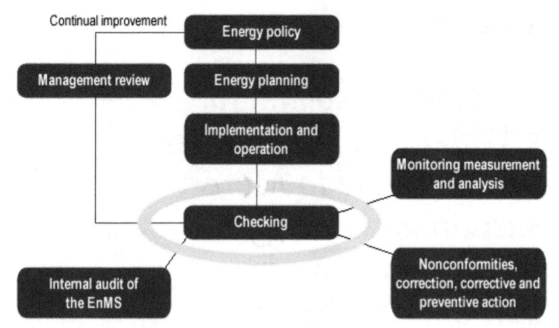

Source: ISO 50001—Energy Management (ISO, 2011); Introduction, pp. v–vi.

establish objectives, targets, and action plans. ISO 50001 applies to activities under the control of the organization, and application of the standard can be tailored to fit specific requirements of the organization, including the complexity of the system, degree of documentation, and resources. The "plan-do-check-act" continual improvement framework is used to incorporate energy management to everyday organizational practices, as the preceding graphic shows.

To simplify mapping of key elements in *The Energy Roadmap* to the clauses in ISO 50001, topics, themes, and progressions are represented in abbreviated fashion as follows: *SM-SD* stands for the strategic management topic area and the strategic direction theme. In instances where a specific characteristic or activity is referenced, the numbers 1, 2, and/or 3 will be appended. For example, *OC-EV23* stands for organizational culture topic, energy vision theme, progressions 2 (integrate) and 3 (optimize). Where no numbers are used, the ISO 50001 clause encompasses all three levels of progression in the theme. Topic areas and theme abbreviations are presented in Table A.1.

Table A.2 shows mapping between *The Energy Roadmap* elements and the ISO 50001 clause. Where ISO 50001 has a subclause denoted by a letter, that letter is listed after the applicable *The Energy Roadmap* element. For example, *OC-EV (a)* in ISO 50001 clause 4.2.1 means that the *The Energy Roadmap* element "Organizational Culture" and topic "Energy Vision" addresses ISO 50001 clause 4.2.1(a).

TABLE A.1   Topic areas and theme abbreviations.

Topic area	Themes
• SM—Strategic Management	• SD—Strategic Direction • FV—Financial Viability • CP—Collaborative Partnerships • TCN—Toward Carbon Neutrality
• OC—Organizational Culture	• EV—Energy Vision • ET—Energy Team • SDA—Staff Development and Alignment
• CO—Communication and Outreach	• CC—Customers and Community • RL—Regulatory and Legislative • MO—Media Outreach • EAG—Environmental Advocacy Groups • WS—Water Sector
• DSM—Demand-Side Management	• ECB—Electricity Costs and Billing • PMC—Power Measurement and Control • EM—Energy Management • SC—Source Control
• EG—Energy Generation	• S—Strategy • EWW—Energy from Water and Wastewater • SES—Supplemental Energy Sources • REC—Renewable Energy Certificates
• IF—Innovating for the Future	• RD—Research and Development • RM—Risk Management • AT—Alternative Technologies • AM—Alternative Management

TABLE A.2   *The Energy Roadmap* crosswalk with ISO 50001—
Energy Management.

ISO 50001 clause	Criteria	*The Energy Roadmap* topic/theme/progression
4.1	General requirements	SM-SD
		OC-EV
4.2	Management responsibility	No clauses
4.2.1	Top management	OC-EV (a)
		OC-ET (b)
		OC-SD (c)
		SM-FV (c)
		SM-SD (d, f, g)
		OC-EV2 (e, i)
		SM-FV2 (h)
		SM SD3 (j)
4.2.2	Management representative	OC-ET12 (a, b)
		OC-ET2 (c, d)
		SM-SD2 (e)
		OC-ET1 (f)
		SM-SD3 (g)
		OC-SDA (h)
4.3	Energy policy	SM-SD (a, h)
		DSM-PMC (b)
		SM-FV (c, f)
		OC-SDA3 (c)
		SM-CP (d)
		SM-SD3 (e)
		EG-REC (f)
		OC-ET23 (g)
4.4	Energy planning	No clauses
4.4.1	General	SM-SD
4.4.2	Legal requirements and other requirements	CO-RL
		EG-S2

TABLE A.2   *The Energy Roadmap* crosswalk with ISO 50001—
Energy Management. (*Continued*)

ISO 50001 clause	Criteria	The Energy Roadmap topic/theme/progression
4.4.3	Energy review	DSM-ECB12 (a)
		DSM-EM1 (a)
		EG-EWW (a)
		EG-SES (a)
		DSM-PMC (b)
		DSM-EM (c)
4.4.4	Energy baseline	DSM-PMC
4.4.5	Energy performance indicators	DSM-PMC
		EG-S
4.4.6	Energy objectives, energy targets, and energy management action plans	DSM-PMC
		DSM-EM
		SM-SD1
		EG-S
4.5	Implementation and operation	No clauses
4.5.1	General	DSM-EM2
		EG-EWW23
		EG-SES23
4.5.2	Competence, training, and awareness	OC-SDA (a, b, c, d)
		OC-EV (a)
4.5.3	Communication	OC-EV2 (internal communications)
		OC-EV3 (external communications)
		CO (external communications)
4.5.4	Documentation	SM-SD
4.5.5	Operational control	DSM-PMC (a, b, c)
4.5.6	Design	DSM-EM3
		SM-FV2

TABLE A.2   *The Energy Roadmap* crosswalk with ISO 50001—
Energy Management. (*Continued*)

ISO 50001 clause	Criteria	*The Energy Roadmap* topic/theme/progression
4.5.7	Procurement of energy services, products, equipment, and energy	SM-CP EG-REC
4.6	Checking	No clauses
4.6.1	Monitoring, measurement, and analysis	DSM-PMC (a, b, c, d, e)
4.6.2	Evaluation of compliance with legal requirements and other requirements	EG-S2 CO-RL
4.6.3	Internal audit of the EnMS*	Procedural aspect of ISO standard not specifically covered by the guidance document
4.6.4	Nonconformities, correction, corrective action, and preventive action	Procedural aspect of ISO standard not specifically covered by the guidance document
4.6.5	Control of records	Procedural aspect of ISO standard not specifically covered by the guidance document
4.7	Management review	No clauses
4.7.1	General	SM OC
4.7.2	Input to management review	Procedural aspect of ISO standard not specifically covered by the guidance document
4.7.3	Output from management Review	Procedural aspect of ISO standard not specifically covered by the guidance document

*EnMS = energy management system.

Index

CPSIA information can be obtained
at www.ICGtesting.com
Printed in the USA
LVHW062116050720
659558LV00008B/17